Complete Dentures — Problem Solving

Complete Dentures — Problem Solving

by

Daryll Jagger,[*] BDS, MSc, FDSRCS (Eng)

Alan Harrison,[†] TD, QHDS, BDS PhD, FDS RCS (Eng)

[*] *Lecturer in Restorative Dentistry, University of Bristol;
Staff Dental Surgeon, The United Bristol Healthcare NHS Trust*

[†] *Professor of Dental Care of the Elderly, University of Bristol;
Consultant in Restorative Dentistry, The United Bristol Healthcare NHS Trust*

1999

Published by the British Dental Association
64 Wimpole Street, London W1M 8AL

© British Dental Journal 1999

All rights reserved. No part of this publication may be reproduced, stored in a retrieval system, or transmitted in any form or by any means, electronic, mechanical, photocopying, recording or otherwise, without either the permission of the publishers or a licence permitting restricted copying in the United Kingdom issued by the Copyright Licensing Agency Ltd, 90 Tottenham Court Road, London W1P 9HE

ISBN 0 904588 57 2

Printed and bound by
Dennis Barber Graphics and Print, Lowestoft, Suffolk

Preface

The provision of complete dentures can be challenging, time consuming and not always successful. Many of the problems can be minimised or avoided by good communication between the dentist and the patient and dentist and dental technician from the outset of treatment.

There should be clear agreement between the patient and the dentist as to the intended outcome of the treatment and it should be planned accordingly. Unreasonable demands and expectations may compromise the success of the denture and should be made clear to the patient. A suitably worded entry in the clinical file, countersigned by the patient, to confirm the discussion and clinical concern may sometimes be appropriate.

The key to problem solving is possession of the facts. A comprehensive medical and dental history and thorough clinical examination is therefore essential and should guide the dentist to a solution. There is a considerable likelihood that denture faults will cause an overlap of complaints. For example, a loose denture may traumatise the tissues and also cause pain. Solving one problem may only create another; the repositioning of anterior teeth in an attempt to satisfy aesthetic demands can result in an unstable denture and a complaint of looseness.

This book is aimed principally at the general dental practitioner but will also be of value to the senior undergraduate. It is written in an informal style and has brought together information not otherwise readily available. Common complaints relating to complete dentures are analysed and a systematic plan to correct the problems is put forward. It is, therefore, an aid to problem solving. The common complaints as they occur in clinical practice are presented, and are accompanied by suggested corrections which can be made to the dentures. Specialised techniques and important issues, such as the prevention of cross infection and the recently introduced Medical Devices Directive 93/42/EEC, are included in detailed appendices.

No attempt is made to cover the more general aspects of the clinical and laboratory stages of denture construction which are available in a range of standard prosthetic text books.

Daryll C. Jagger and Alan Harrison
Bristol, March 1999

Acknowledgements

We would like to express our gratitude to Mr C. Bell for helpful guidance and criticism, Mr R. G. Jagger for his advice on Chapter 6, Mr C. M. Woodhead for the use of Figures 1.4-1.6, 3.3, A2.1 and A4.1-4.4 and Dr J. W. Eveson for Figure 2.6. The basis of some of the contents was a number of articles originally published in the *British Dental Journal* (Chapter 5 and Appendices 5, 6 and 8) and the *Journal of Primary Dental Care* (Chapter 3). We would like to thank the Editors, Mr M. Grace and Professor E. Renson, who kindly agreed that some of this material could be reproduced.

Contents

Foreword		v
Preface		vi
Acknowledgements		vi
1	The loose denture	1
2	The painful denture	5
3	The broken denture	9
4	Aesthetic denture problems	13
5	The dirty denture	17
6	The 'problem' denture	21
Appendix 1	Increased gag reflex	25
Appendix 2	The neutral zone	27
Appendix 3	Cuspless teeth	31
Appendix 4	Copy dentures	33
Appendix 5	Denture fixatives	37
Appendix 6	Soft lining materials	39
Appendix 7	Medical Devices Directive	43
Appendix 8	Cross-infection control	45
Index		47

complete dentures

1 The loose denture

The aim of this chapter is to discuss the likely causes of looseness of dentures and to make recommendations on its management.

Perhaps the most common complaint with dentures is that of looseness. This is more commonly associated with the lower denture and usually brought to the dentist's attention either soon after the dentures are fitted or following a period of successful wear when the dentures are nearing the end of a useful life. The aim of this chapter is to discuss the likely causes of looseness of dentures and to make recommendations on its management.

The causes of looseness can be divided into those attributed to the denture, for example underextension or overextension of flanges, and those attributed to the denture wearer, such as atrophic alveolar ridges or neuromuscular disorders. There can, of course, be a combination of denture design faults and problems attributable to the patient.

Problems with the denture

Impression surface

a. Poor impression. If the impression surface of the denture does not closely conform to that of the denture-bearing tissues, trapped air and saliva will fill the space and the retention will be reduced. This can be avoided by ensuring that:
- a special tray is used and that it is carefully adapted to the underlying tissues such that it supports a uniform thickness of impression material
- a suitable impression material, such as zinc oxide eugenol, light or medium bodied silicone, or alginate is chosen for the working impression
- the borders of the material are adequately supported (if necessary with an impression tracing compound)
- the impression material is poured as soon as possible to avoid distortion or suitably prepared for transportation to the dental laboratory.

If the impression surface of the denture is at fault a reline may be necessary to correct any deficiencies.

b. Damaged working model. Considerable time and effort can go into making high quality preliminary and final impressions and the resultant working models should be carefully handled since their accuracy will determine the final retention and stability of the dentures. If damage to the working model is likely on removal of the impression, the technician should consider sectional removal of the special tray. The technician should trim the models but ensure that the reproduction of the depth and width of the borders of the impressions and hence the sulci are retained (fig. 1.1). Narrow ridges on lower models are especially susceptible to damage and, if fractured and poorly repaired using adhesive in the laboratory, can ruin the accuracy of the impression surface of the finished denture.

c. Warped denture. Dimensional change during processing can be avoided by using a careful controlled conventional processing technique to fully polymerise the poly(methyl methacrylate). When the laboratory stages are completed the denture should be kept moist since drying may cause a change in the shape.

d. Excessive palatal relief chamber. Relief chambers are rarely necessary. However, in situations where the mucosa is more compressible over the ridges than the midline or where there is a prominent palatal torus it may be necessary to incorporate one within the denture base to prevent excessive flexing of the denture. If the relief chamber is extensive the retentive seal can be broken more easily and the denture will be loose. Epithelial proliferation beneath a palatal relief chamber can occur (fig. 1.2).

e. Absent or deficient post-dam. An inadequate, even-depth, post-dam in the form of a narrow line is commonly incorporated at the posterior (often underextended) border of the upper denture. To function efficiently the post-dam must be placed at the junction of the hard and soft palate on compressible tissue and extended laterally to the mucosa overlying the hamular notches. A cupid's bow design is preferable (fig. 1.3). Ideally the post-dam

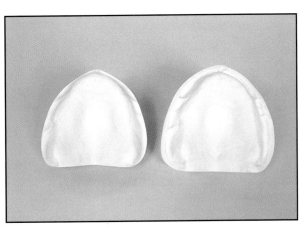

Fig. 1.1 Examples of the same model, on the right, correctly trimmed to ensure that the land adjacent to the sulcus is maintained and on the left, overtrimmed illustrating loss of sulcus form. This commonly results in knife edged underextended peripheries on the finished denture.

complete dentures

should be carved on the working model by the dentist since the technician cannot be aware of the local anatomy and tissue compressibility. Additions to an existing ineffective post-dam can be made at the chairside with an autopolymerising acrylic resin designed for intra oral use such as Tokuso (Tokuyama Corporation, Tokyo, Japan). Care should be taken not to extend the post-dam excessively beyond the junction of the hard and soft palate as this, together with other factors (for example; tongue cramping due to a lingual position of the posterior teeth, thickened tuberosities, incorrect occlusal plane) can cause a sensation of nausea and an increased gag reflex (see Appendix 1). An absent post-dam should be corrected in the laboratory following a reline impression.

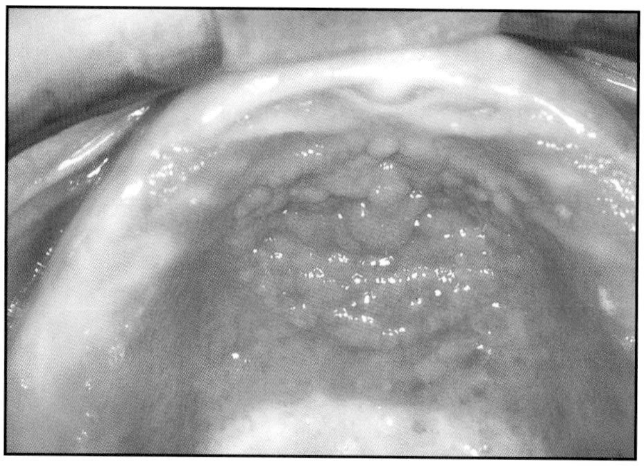

Polished surfaces

It is essential to produce an effective border seal to prevent the ingress of saliva and air and a corresponding reduction in retention.

a. Underextended borders in depth and width. The denture should fill and form a seal, both with the full functional depth and width of the sulci and at the posterior border. During denture construction this can be achieved with the use of:

- carefully trimmed special trays (fig. 1.4)
- adequate border moulding of the peripheries with impression tracing compound (fig. 1.5)
- a suitable impression technique to record the functional depth and width of the sulci and preservation of this detail on the working model and subsequent denture (fig. 1.6).

Careful examination of the peripheries by direct vision, coupled with the diagnostic addition of tracing compound should reveal areas of underextension. In particular, attention should be paid to the distobuccal areas of the maxillary tuberosities and hamular notches and the distolingual sulci and the retromolar pads in the lower arch. Areas of under extension in a completed denture should be corrected by careful border moulding with tracing compound and a subsequent reline impression.

b. Overextended peripheries in depth and width. As for underextended peripheries, they should be identified by direct vision utilising a soft disclosing cream or wax if necessary. Small areas can often be corrected by careful trimming and polishing however for gross overextension it may be necessary to reduce the peripheries short of the functional sulci and to border mould with tracing compound followed by a reline impression.

c. Polished surfaces not in the neutral zone. The neutral zone is reported to be the area of minimal conflict between the lips cheeks and tongue in which the denture should be most stable. If the polished surfaces of the denture or teeth lie outside the neutral zone they are more likely to be displaced by the oral musculature and result in a loose denture. In these circumstances the denture contours should either be modified or, if necessary, remade using the neutral zone technique (see Appendix 2). Plumping of the

Fig. 1.2 Epithelial proliferation in the area defined by a palatal relief chamber with superimposed denture induced stomatitis.

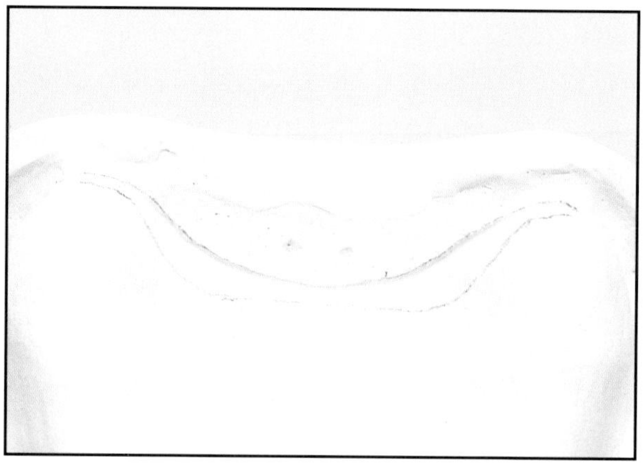

Fig. 1.3 A cupid's bow post dam carved on a model demonstrating the relationship of the (exaggerated) fovea palatinae. Note the gradual chamfer deepening towards the posterior aspect.

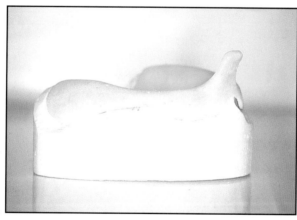

Fig. 1.4 A special tray on the primary model correctly trimmed to allow for border moulding.

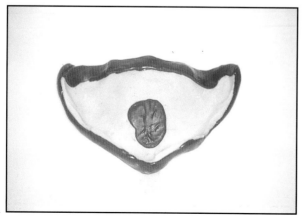

Fig. 1.5 An example of border moulding using greenstick tracing compound.

complete dentures

upper anterior flange or excessive prominence of the anterior teeth in an attempt to satisfy the patient's aesthetic needs can reduce the fit and stability of the denture and often a balance has to be achieved between aesthetics and function.

Occlusal surface

Forces which tend to destabilise the denture during chewing and speech are often associated with occlusal irregularities.

a. Intercuspal position not coincident with retruded contact position. A precentric (no tooth contact) record should be taken, the dentures mounted on an articulator and the dentures adjusted to produce a balanced occlusion with freedom of movement from retruded contact position to intercuspal position. The procedure is carried out in two stages; the correction of the centric occlusion at the correct centric relation and vertical dimension and then the development of eccentric balancing contacts. To correct centric occlusion (stage 1):

- if a cusp is high in both centric and eccentric positions reduce the cusp
- if a cusp is high in centric but not in eccentric occlusion deepen the fossa
- if a cusp is high in eccentric occlusion only leave for stage 2.

To develop eccentric contacts selective grinding is carried out according to the BULL rule (reduction of buccal upper/lingual lower) (stage 2):

- no reduction of the maxillary lingual cusp
- no reduction of the mandibular buccal cusp
- no deepening of the fossae on any tooth.

The occlusion should be adjusted on the working side by reducing the inner inclines of the maxillary buccal cusps and mandibular lingual cusps and on the balancing side by reducing the inner inclines of the maxillary palatal or mandibular buccal cusps. In protrusion the occlusion should be adjusted by grinding the mesial inclines of the mandibular buccal cusp, the distal incline of the maxillary lingual cusp, the lingual surfaces of the upper anterior teeth and the labial surfaces of the incisal edges of the mandibular teeth.

b. Premature occlusal contact. Minor errors can be corrected at the chairside by selective occlusal adjustment. The denture bases should be securely in place and thin articulating paper used to mark and record the premature contacts. For more substantial errors a precentric record should be taken (ensuring that the teeth do not come into contact) (fig. 1.7), the dentures remounted on an articulator and the occlusion adjusted in the laboratory. For gross discrepancies it may be necessary to remove the teeth, replace with wax occlusal rims and re-record the centric jaw relationship. The teeth should then be reset for a new wax trial.

c. Locked occlusion. For some individuals, especially the elderly who may be used to worn (flat) occlusal surfaces, it is often difficult to adapt to a cusp fossa occlusal relationship or a deep anterior overbite. In these cases it may be necessary to selectively grind the occlusal surfaces to allow a freedom of movement particularly in lateral excursions, and to reduce the overbite. The use of non-anatomic (cuspless) teeth should be considered (see Appendix 3).

Moderate wear of the occlusal surfaces usually allows free jaw movements and should not cause a complaint of looseness. However excessive wear, with a loss of occlusal vertical dimension can result in a postural Class III jaw relationship with subsequent locking of the lower anterior teeth in front of the uppers. This problem can be rectified by re-establishing an appropriate vertical dimension and occlusal relationship by adding autopolymerising acrylic resin to the lower (usually) occlusal surfaces of the posterior teeth, in stages, to allow for a period of adaptation to the increase in height.

d. Occlusal plane incorrect. If the occlusal plane is not parallel to the ridge, moments of force can be produced which tend to displace the denture. If the occlusal plane of the lower denture is too high, the tongue will not be able to rest on the teeth and will tend to displace the denture rather than help secure it. In this case, if the occlusal plane on the upper denture is correct, the lower occlusal plane will have been raised at the expense of the freeway space and it will be necessary to remake the lower denture at the correct occlusal vertical dimension.

Further reading

Basker R M, Davenport J C, Tomlin H R. *Prosthetic treatment of the edentulous patient* 3rd edn. Basingstoke: Macmillan Press, 1992.

Grant A A, Heath J R, McCord J F. *Complete prosthodontics. Problems, diagnosis and management.* London: Mosby Year Book Europe Ltd, 1994.

Hopkins R. *A colour atlas of preprosthetic oral surgery.* London: Wolfe Medical, 1987.

MacGregor A R. *Clinical dental prosthetics* 3rd edn. London: Wright, 1989.

Pankhurst C L, Dunne S M, Rogers J O. Restorative dentistry in the patient with dry mouth. Part 2. *Dent Update* 1996; **23**: 110-114.

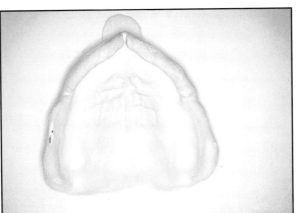

Fig. 1.6 A working impression demonstrating good surface detail and reproduction of the functional depth and width of the sulci.

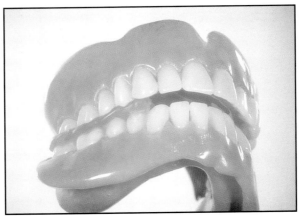

Fig. 1.7 A wax precentric record registered with the teeth slightly apart in order to avoid tooth contact which may guide the teeth into the original occlusion or displace the dentures.

complete dentures

Problems with the denture wearer

Poor neuromuscular control

There are a number of neuromuscular disorders, eg Parkinson's disease and cerebrovascular accidents (strokes), which can affect the prognosis of complete dentures due to a diminished muscular control. It is essential to provide dentures with optimum retention and stability. The use of non-anatomic (cuspless) teeth together with a neutral zone technique should be considered. It may be desirable to use a copy denture technique (see Appendix 4) to copy features of old dentures and to incorporate selective modifications to improve them where appropriate. For some individuals, retention is a major difficulty and the regular use of denture fixatives can be an option (see Appendix 5).

Unstable foundations

If the foundations for the denture are unfavourable, eg an upper anterior flabby ridge (fig. 1.8), atrophic lower ridge (fig. 1.9), high fraenal attachments (figs 1.10 and 1.11) or prominent mylohyoid ridges, it may be necessary to consider the use of special impression techniques or even preprosthetic surgery to produce an improved anatomy and the optimum denture. For the upper anterior flabby ridge, it is important that the impression does not displace the flabby tissue. If this tissue is displaced the denture may be well adapted when seated by occlusal pressure but is likely to be displaced by the elastic recoil of the displaced tissue when the teeth are apart. The fibrous tissue will be further damaged by the pumping action and the movement of the denture caused by tooth contact. If there is only minimal displacement of the flabby ridge then it is possible to use a spaced special tray with a low viscosity material such as impression plaster or low viscosity alginate. With marked displacement it is better to use a two-part impression technique. A close fitting special tray, with a window over the flabby area is used to take an impression of the undisplaced mucosa in zinc oxide eugenol impression paste. The impression is checked, replaced in the mouth and an impression taken of the flabby ridge by adding impression plaster to the window area.

The atrophic mandibular ridge is prevalent because of the process of resorbtion which results in approximately four times as much bone loss in the mandible as in the maxilla. The common complaint of a loose lower denture often is the result of the combination of poor anatomy and an underextended denture base. Problems with retention of upper dentures are less frequent but can be very difficult to treat successfully if the maxilla is very atrophic.

Inadequate saliva

Adequate saliva is essential to aid retention in the denture wearer. For individuals with xerostomia (eg Sjögren's syndrome) retention can be a major problem. Dentures should be provided which maximise retention and stability together with the provision of an artificial saliva substitute where appropriate. The use of denture fixatives is an option.

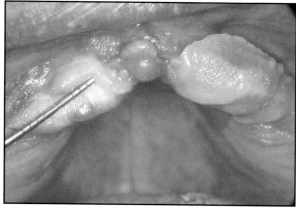

Fig. 1.8 A mobile, flabby anterior maxillary ridge which is easily compressed.

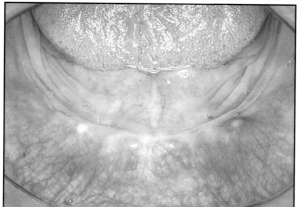

Fig. 1.9 An atrophic mandibular ridge. An amalgam tattoo is present in the left premolar region.

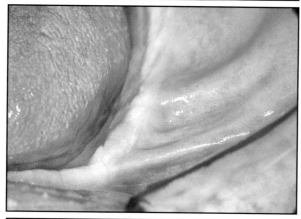

Fig. 1.10 A high buccal frenal attachment near the crest of the edentulous mandibular ridge.

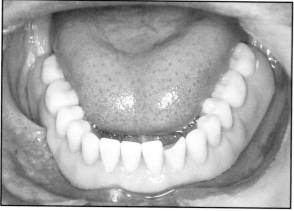

Fig. 1.11 A lower complete denture overlying an unfavourable buccal frenal attachment which could lead to denture displacement.

complete dentures

2 The painful denture

Pain is a common complaint associated with dentures, and as with looseness can be brought to the dentist's attention soon after the dentures are fitted or after a period of successful wear. It is more commonly associated with the lower denture. The aim of this chapter is to discuss the likely causes of pain associated with dentures and make recommendations on its management. The causes of pain can be divided into those attributed to the denture, for example overextension of flanges, and those attributed to the denture wearer, such as unfavourable anatomy of the denture bearing tissues. Alternatively the causes can be classified according to the location of the pain, ie localised or generalised.

Problems associated with the denture

Impression surface

a. Damaged fit surface. Blebs or spurs of acrylic resin on the fit surface of a denture can lead to localised pain under the denture (fig. 2.1). They can be the result of a faulty impression, damaged model, poor processing techniques or a combination of these factors. They can be detected by simple touch or by dragging a piece of cotton wool or gauze over the fit surface and the denture should be adjusted accordingly.

b. Extension of the borders of the denture into deep undercuts. Extension of the borders of dentures into deep undercuts will cause localised pain on insertion and removal of the denture. This is a problem frequently found in the upper anterior labial areas and distal to the tuberosities. Areas of erythema and ulceration should be identified by direct vision or with the aid of a pressure indication paste or spray. For obvious ulceration, the lesion should be gently wiped dry with cotton gauze, and pressure indicating paste applied, with a small brush, to the lesion. The denture should be reinserted in the mouth, gentle finger pressure applied, the denture removed and examined on the fit surface, for areas of paste corresponding to areas of mucosal ulceration. The denture should be adjusted in these areas, replaced in the mouth and the procedure repeated if necessary.

c. Inadequate relief. Appropriate relief of the denture is required over defined areas of incompressible mucosa, for example bony prominences, and where the mucosa is atrophic, easily traumatised and unable to withstand high loads. Lack of, or inadequate, relief in these areas can produce localised or generalised pain. A permanent soft lining material (on a lower denture) may be helpful in absorbing and disseminating some of the occlusal loads in these circumstances (see Appendix 6).

Inadequate relief around muscle attachments, particularly the labial and buccal frenae will produce localised pain. The denture should be adjusted to alleviate the trauma but care should be taken not to remove more material than necessary, otherwise the border seal will be disrupted. Preprosthetic surgery should be considered for muscle attachments which compromise the success of the denture.

d. Deep post-dam. An excessively deep and/or overextended post-dam can be traumatic and produce localised pain together with a more generalised pharyngeal pain (fig. 2.2). The denture should be adjusted to relieve the discomfort. If gross adjustments are made, a reline impression may be necessary in order to ensure close adaptation of the fit surface.

The aim of this chapter is to discuss the likely causes of pain associated with dentures and make recommendations on its management

Fig. 2.1 A complete upper denture with a clear acrylic resin palate with an obvious bleb on the fit surface.

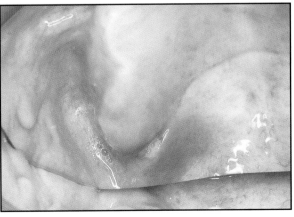

Fig. 2.2 A traumatic ulcer distal to the tuberosity related to the posterior border of the denture.

complete dentures

Polished surfaces

a. Overextension of the borders in depth and width. Examination of the peripheries by direct vision coupled with careful manipulation of the oral musculature will reveal areas of overextension. Overextension is often associated with looseness, especially in the lower denture, although it can also cause pain, and is usually in the lower anterior and posterior lingual regions. For the upper denture, where retention is often better, overextension is more likely to cause pain (fig. 2.3). The denture should be adjusted to correct the overextension.

b. Polished surfaces and teeth not in the neutral zone. Incorrect positioning of the teeth can result in cheek, lip or tongue biting with obvious discomfort (figs 2.4 and 2.5). Cheek and lip biting tend to be associated with insufficient buccal and incisal horizontal overjet which can be corrected by adding wax to the facial surface of the upper teeth as a trial procedure and then resetting the teeth accordingly. Tongue biting is usually due to a combination of the posterior teeth being placed in a lingual position, together with sharpness of the cusps. The teeth should be either reduced on the lingual aspect and smoothed or, if the positioning error is more obvious, reset in the correct position.

Occlusal surfaces

a. Premature occlusal contact. Pain tends to be localised to the ridge in the region of a premature occlusal contact, but can be transferred to distant sites, due to movement of the denture and trapping of the mucosa beneath the base. Minor errors can be corrected at the chairside by selective occlusal adjustment. The denture bases should be securely in place and thin articulating paper used to mark and record the premature contacts. For more substantial errors a precentric record should be taken (ensuring that the teeth do not come into contact), the dentures remounted on an articulator and the occlusion adjusted in the laboratory (see Chapter 1). For gross discrepancies it may be necessary to remove the teeth, replace with wax occlusal rims and re-record the centric jaw relationship. The teeth should then be reset in wax for a further trial in the mouth.

b. Intercuspal position not coincident with retruded contact position. A precentric record should be taken, the dentures mounted on an articulator and the dentures adjusted to produce a balanced occlusion with freedom of movement from retruded contact position to intercuspal position.

c. Locked occlusion. For some individuals, especially the elderly who may be used to worn (flat) occlusal surfaces, it is often difficult to adapt to a cusp fossa occlusal relationship or a deep anterior overbite. In these cases it may be necessary to selectively grind the occlusal surfaces to allow a freedom of movement particularly in lateral excursions, and to reduce the overbite. With an excessive overbite pain may be experienced around the incisal papilla as the upper denture pivots around this fulcrum.

d. Increased occlusal vertical dimension. With a reduction, or complete loss, of freeway space pain is usually generalised and most obvious in the lower arch and there may also be discomfort in the muscles of mastication. The dentures should be modified, usually by adjusting the occlusal surfaces if the error is small, to provide adequate freeway space. More substantial errors will require a remake at the correct occlusal vertical dimension.

Factors related to the denture wearer

a. Unfavourable anatomy. In the mandible, gradual resorption of the alveolar ridge with

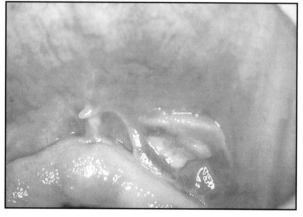

Fig. 2.3 An example of tissue ulceration and proliferation associated with localised overextension of the denture border.

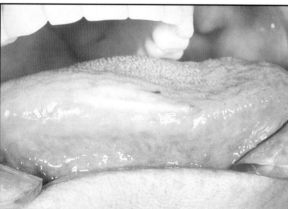

Fig. 2.4 A white lesion on the lateral border of the tongue in a denture wearer who complained of persistent tongue biting.

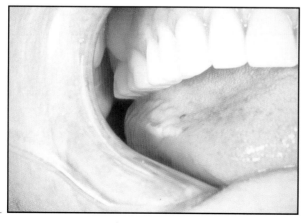

Fig. 2.5 The indentations on the lesion correspond to the occlusal form of the posterior teeth of the upper complete denture and confirm the traumatic origin.

age leads to a reduction in the denture-bearing area and often a knife edged atrophic ridge. The overlying mucosa can be thin and easily traumatised by the overlying hard denture base, particularly in those individuals with clenching or grinding habits. A soft lining (see Appendix 6) may be helpful in absorbing and disseminating some of the occlusal loads in these circumstances. In some individuals excessive resorption may lead to the mental nerve lying on the surface of the residual ridge. The nerve can be trapped between the hard denture base and the bone causing a sharp, shooting pain. Inclusion of appropriate relief on the master cast together with a soft lining may help to reduce the pressure on the nerve.

Mucosa overlying prominent maxillary and mandibular tori or mylohyoid ridges, is often thin, and can be easily traumatised. A soft lining may also be used in these circumstances, however although this is often advocated in textbooks it is not a widely adopted treatment option, as it increases the bulk of and weakens the denture, especially in the mid palatal region over maxillary tori. An alternative approach is simply to relieve the denture over the prominent areas.

b. Systemic disease. Buried teeth, roots, cysts, osteodystrophies (eg Paget's disease) and tumours (fig. 2.6) are examples of local lesions which can affect the fit of dentures or cause discomfort and pain. Although commonly associated with local, easily manageable dental problems, the pain may on occasions be a reflection of a serious underlying condition.

Aphthous ulceration, lichen planus, vesicular or bullous disorders (eg pemphigus, benign mucous membrane pemphigoid), herpes zoster, haematological deficiencies (eg serum and red cell folate, serum iron, and serum vitamin B12) and carcinoma are all examples of generalised pathology which can cause discomfort and pain under dentures. It is important therefore to exclude systemic disease as a possible factor. The cause of persistent pain under dentures should be identified and not evaded by repeated eases.

c. Xerostomia. A reduced saliva flow, possibly as a result of degenerative changes in the salivary glands, radiotherapy or drug therapy can result in loose dentures which can cause discomfort and soreness. If pharmacotherapy is the culprit it may be possible for the patient's general medical practitioner to alter the regime. The use of artificial saliva should be considered.

d. Allergy. Allergy to poly (methyl methacrylate) and methyl methacrylate monomer is rare and patch testing of the individual is necessary to diagnose a true allergy. It is more likely that mucosal inflammation and/or ulceration is due to chemical irritation to the direct contact with excessive methyl methacrylate monomer (fig. 2.7). The denture base should be analysed to assess the residual monomer content which should be less than 1.0%. This can only be performed in a specialist centre using for example, gas liquid chromatography. If there is an excess of residual monomer the denture should be returned to the laboratory and subjected to a further controlled heat curing cycle.

e. Psychological. In some cases the cause of the pain is unclear and there can be an underlying psychological problem (see Chapter 6).

Further reading

Austin A T, Basker R M. The level of residual monomer in acrylic denture base materials with particular reference to a modified method of analysis. *Br Dent J* 1980; **149:** 281-286.

Basker R M, Davenport J C, Tomlin H R. *Prosthetic treatment of the edentulous patient.* 3rd edn. Basingstoke: Macmillan Press, 1992.

Grant A A, Heath J R, McCord J F. *Complete prosthodontics. Problems, diagnosis and management.* London: Mosby Year Book Europe Ltd, 1994.

Zografakis M A, Harrison A, Huggett R. Measurement of residual monomer in denture base material: studies on variations in methodology using gas liquid chromatography. *Eur J Prosthodont Restor Dent* 1994; **2:** 101-107.

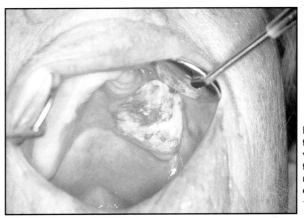

Fig. 2.6 A tumour in the upper arch which was only evident to the clinician on removal of the denture.

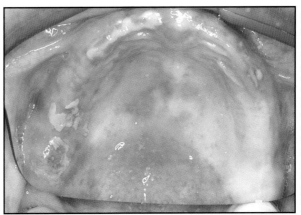

Fig. 2.7 Inflammation and ulceration of a maxillary arch, predominantly affecting the ridge, which occurred within 48 hours of insertion of a new denture. Gas liquid chromatography analysis of the denture base material demonstrated a residual monomer level in excess of 3%.

complete dentures

3 The broken denture

The material most commonly employed in the construction of dentures is poly (methyl methacrylate) (PMMA) acrylic resin. Although satisfactory aesthetics can be achieved, in terms of its mechanical properties (eg strength) it is still far from ideal in fulfilling the requirements of a prosthesis. The fracture of acrylic resin dentures is an unresolved problem causing inconvenience to both the dentist and denture wearer. A multiplicity of factors may be responsible for the ultimate failure of a denture and failure is not necessarily due entirely to the intrinsic properties of the denture base material. The cause of a fracture must be identified and corrected before a denture is repaired or replaced, otherwise the denture is likely to fracture again for the same reason.

The aim of this chapter is to discuss factors predisposing a denture to fracture, to make recommendations as to how these problems can be minimised and to provide an update on methods of reinforcement of denture base materials.

Fracture of dentures

Up until 1990, 34.9 million repairs to dentures had been carried out since the National Health Service began in 1948, at a cost of £67.8 million (not allowing for variations in monetary value). At present the Dental Practice Board spends approximately £7 million annually to repair about 0.8 million dentures. The overall cost of repairing broken dentures is no doubt even greater since these figures do not take into account repairs undertaken in hospitals, the community service or privately.

Despite the severity of the problem there have been few reports on the prevalence of fractured dentures. A survey in 1969 reported that 63% of dentures had broken within three years of their provision, there being a greater proportion of partial dentures than complete dentures. Another survey of the prevalence of type of fracture conducted by distribution of a questionnaire to three laboratories, reported that 33% of the repairs carried out were due to debonded/detached teeth and 29% were repairs to midline fractures, more commonly seen in upper complete dentures. The remaining 38% were other types of fractures, the majority of which were repairs to upper partial dentures eg detachment of acrylic resin saddles from metal in metal based dentures and fracture of connectors in all acrylic resin partial dentures.

Fractures in dentures result from two different types of forces, namely, flexural fatigue and impact. Flexural fatigue occurs after repeated flexing of a material and is a mode of fracture whereby a structure eventually fails after being repeatedly subjected to loads that are so small that one application apparently does nothing detrimental to the component. This type of failure can be explained by the development of microscopic cracks in areas of stress concentration. With continued loadings, these cracks fuse to an ever growing fissure that insidiously weakens the material. Catastrophic failure results from a final loading cycle that exceeds the mechanical capacity of the remaining sound portion of the material. The midline fracture in a denture is often due to flexural fatigue.

Impact failures normally occur out of the mouth as a result of a sudden blow to the denture or accidental dropping during cleaning, or when coughing or sneezing.

Factors predisposing a denture to fracture

Deformation of dentures

Any factor which increases the deformation of a denture base will predispose the denture to fracture. The functional deformation of complete dentures has been studied using strain gauges and the functional strain in maxillary dentures and the direction and degree of maximum and minimum functional strain in maxillary dentures have also been reported. A denture base in function exhibits considerable flexure. A complete denture is likely to have flexed several million times during its functional life since it has been estimated that it flexes 500,000 times a year.

Deformation of complete maxillary dentures can occur when the teeth contact, for example during chewing, swallowing and clenching. Ideally dentures should fit properly and deform minimally. The optimum situation would be for there to be the same degree of functional deformation in the denture base material as in the supporting tissues. However because functional changes in the tissues are small the denture bases are generally made rigid. Deformation and/or movement of the dentures during function will affect both the supporting

> **The aim of this chapter is to discuss factors predisposing a denture to fracture, to make recommendations as to how these problems can be minimised and to provide an update on methods of reinforcement of denture base materials.**

complete dentures

tissues and the denture base itself. The type of denture base deformation appears to be defined by the anatomy of the supporting tissues with high ridge bases exhibiting torsion deformation while compression (inward movement) occurs with flat ridges, and also by occlusal design with horizontal deformation increasing with the cusp angle during mastication. Occlusal surface area has little effect on deformation.

Stress concentrators

Additional factors which form areas of stress concentration and can predispose a denture to fracture include a large fraenal notch (fig. 3.1), midline diastema, foreign particles, gas inclusions and surface irregularities such as scratches and abrupt contour changes. Stress concentration can occur around the pins of porcelain teeth or metal strengtheners.

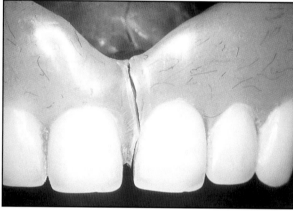

Fig. 3.1 A midline fracture through the franal notch and midline diastema of a complete upper denture.

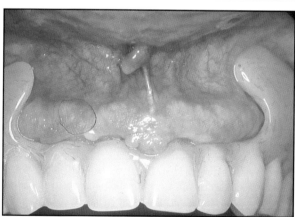

Fig. 3.2 The lack of an upper anterior flange in this complete denture weakens the overall structure of the denture base and may increase the risk of fracture.

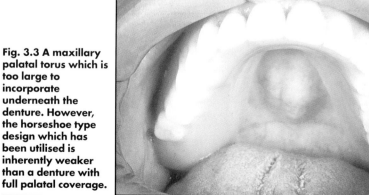

Fig. 3.3 A maxillary palatal torus which is too large to incorporate underneath the denture. However, the horseshoe type design which has been utilised is inherently weaker than a denture with full palatal coverage.

Thin or underextended flanges and absent flanges

Dentures with thin or underextended flanges tend to be prone to additional flexure as are open face (flangeless) dentures (fig. 3.2). If a decision is taken to design a denture without an anterior flange, compensatory measures should be considered, for example, the inclusion of a metal palate or the use of an alternative denture base material, such as a high impact acrylic resin (discussed later in the chapter).

Poor fit

As a result of alveolar resorbtion, a denture will become poorly adapted to the underlying denture bearing tissues and therefore prone to excessive flexure and subsequent fracture during function. This situation can be rectified by ensuring the denture is relined or rebased before the denture is of sufficiently poor fit to create a problem.

Lack of adequate relief

In the case of a bony prominence, such as a maxillary torus, if it is of modest size it may be possible to incorporate it under the fitting surface of a normally extended denture. It is essential that adequate relief is provided in the denture together with appropriate postdamming to ensure satisfactory retention. This will eliminate the need for a horseshoe type design, which will have reduced retention, (unless there are substantial alveolar ridges), and is more likely to flex and fracture (fig. 3.3).

Tooth wear

An occlusion which is locked, worn or causing wedging may predispose a denture to fracture. A wedging action can occur in old complete dentures because of the pattern of alveolar resorbtion and the wear of the artificial teeth leading to an upward and outward loading of the maxillary denture. Heavy occlusal loads can occur in bruxism, (or when a complete denture opposes a natural dentition), and care must be taken to produce an even and balanced occlusion to reduce the risk of fracture.

Previous repairs

Previously repaired dentures are prone to repeated fracture since they are only approximately 80% of the strength of the original resin and will almost certainly break again if the reason for failure has not been corrected. Autopolymerising resin is still the most commonly used to repair dentures because of the rapid and easy processing technique.

The contour of the repair surface of fractured dentures affects the strength of the repair. For example, a 45° edge profile and a rounded surface have been reported to provide the best strength of the repair joint. The strength of an autopolymerising resin repair has been shown to be time dependent. The strength increases during the first week after the repair, and can also be improved by hydroflask curing and microwave irradiation.

Reinforcement of dentures

Over the years various approaches to strengthening acrylic resin have been suggested including modifying or reinforcing the resin. Strengthening has been approached through chemical modification to produce graft copolymers of rubber methacrylate (referred to as high impact resins) or by incorporating solid or open mesh metal forms or various types of fibres to provide reinforcement of fracture prone areas.

Chemical modification of PMMA

The addition of rubbers to PMMA produces a resin that consists of a matrix of PMMA within which is dispersed an interpenetrating network of rubber and PMMA. The objective of these 'high impact' resins is self evident in that they absorb greater amounts of energy at higher strain rates before fracture than do standard resins. Also, an increase in fatigue resistance is claimed because in principle a developing crack will accelerate in PMMA but decelerate across the rubber phase.

The development of high impact strength denture base materials using a low molecular weight (15-35,000) butadiene styrene rubber has been discussed in the dental literature. This rubber can be incorporated up to 30% w/v in methyl methacrylate without deleterious effects, such as a marked increase in viscosity, on the handling characteristics. The butadiene styrene also has reactive (acrylate) end groups to facilitate grafting to the PMMA.

A problem with rubber reinforcement of resins is that whilst the impact strength can be improved it is at the expense of the Young's modulus, producing a denture base with increased impact strength but which may be too flexible.

This method of reinforcement is to date the most successful and widely accepted and is an alternative to the conventional PMMA denture base resin, however the high cost, often up to 20 times that of conventional resin, restricts its routine use. A number of these high impact resins are available and include Colacryl HS 2100 (Bonar Polymers, Newton Aycliffe, UK) and Acron High (Austenal Dental Products, Harrow, UK).

Addition of other fibres to PMMA

Several types of fibres have been added to acrylic denture resins in attempts to improve their physical and mechanical properties. Reinforcement has been achieved through the addition of polyaramid fibres, sapphire whiskers, carbon and glass fibres. The use of carbon and polyaramid fibres is limited because of difficulties in polishing and poor aesthetics. Problems can arise with fibres exposed at the surface of the resin which present a rough surface that cannot be polished and can be uncomfortable for the patient. The use of glass fibres has produced some improvement in mechanical properties of the resin. In recent years the inclusion of oriented or woven ultra high modulus polyethylene fibres (UHMPE) has produced encouraging results, however the technical difficulties, coupled with the additional time required, limits their use. UHMPE beads in both untreated and surface treated form have not been shown to be of any significant value in increasing the strength of the base resin.

Addition of metal strengtheners to PMMA

The use of metal inserts to reinforce denture base resin has been reported to affect the fracture resistance of acrylic resin specimens. Metals can be added in the form of wires, bars, mesh, plates or fillers. The thickness and position of the strengthener within the resin can affect its reinforcing properties. For maximum strength the metal should be placed perpendicular to the anticipated line of stress and fracture and not coincident with that line. The inclusion of metal bars or wires as a means of reinforcement of PMMA often has limited value because the increase of stress concentration may outweigh the benefits.

Attempts to strengthen acrylic resin can result in failure at the resin/strengthener interface. Areas of stress concentration occur around embedded materials and the overall effect can be to weaken rather than strengthen the denture base. In an effort to improve the interface techniques such as sandblasting, silanisation and the use of metal adhesive resins have been used. Failure at the interface between acrylic resin and the reinforcement material is an unresolved problem. For lower complete dentures which have a permanent resilient soft lining the use of a metal strengthener, usually in the form of a cobalt chromium lingual plate, is a reasonable option.

Repair of broken dentures

The cause of the fracture must be identified before the denture is repaired, otherwise repeated fracture is likely. For a simple midline fracture, the two fragments should be assembled and secured in position with sticky wax together with additional reinforcement, for example wooden sticks across the line of fracture (fig. 3.4). If possible, the accuracy of the temporary 'repair' should be checked by seating the denture in the mouth. The denture is then sent to the dental laboratory where any undercuts are blocked out with cotton gauze and dental plaster poured into the impression surface.

Further reading

Beyli M S, von Fraunhofer J A. An analysis of causes of fracture of acrylic resin dentures. *J Prosthet Dent* 1981; **46**: 238-240.

Bowman A J, Manley T R. The elimination of breakages in upper dentures by reinforcement with carbon fibre. *Br Dent J* 1984; **156**: 87-89.

Carlos B, Harrison A. The effect of untreated UHMWPE beads on some properties of acrylic resin denture base material. *J Dent* 1997; **25**: 59-64.

Darbar U R, Huggett R, Harrison A. Denture fracture. A survey. *Br Dent J* 1994; **176**: 342-345.

Glantz P-O, Stafford G D. Clinical deformation of maxillary complete dentures. *J Dent* 1982; **11**: 224-230.

Hargreaves A S. The prevalence of fractured dentures. A survey. *Br Dent J* 1969; **126**: 451-455.

Harrison W M, Stansbury B E. The effect of joint surface contours on the transverse strength of repaired acrylic resin. *J Prosthet Dent* 1970; **23**: 464-472.

Kydd W L. Complete base deformation with varied occlusal tooth form. *J Prosthet Dent* 1956; **6**: 714-718.

Ladizesky N H, Chow T W, Cheng Y Y. Denture base reinforcement using woven polyethylene fiber. *Int J Prosthodont* 1994; **7**: 307-314.

Polyzois G L. Reinforcement of denture acrylic resin. The effect of metal inserts and denture resin type on fracture resistance. *Eur J Prosthodont Restor Dent* 1995; **3**: 275-278.

Rees J S, Huggett R, Harrison A. Finite element analysis of the stress concentrating effect of frenal notches in complete dentures. *Int J Prosthodont* 1990; **3**: 238-240.

Rodford R A. Further development and evaluation of high-impact-strength denture base materials. *J Dent* 1990; **18**: 151-157.

Solnit G S. The effect of methyl methacrylate reinforcement with silane-treated and untreated glass fibers. *J Prosthet Dent* 1991; **66**: 310-314.

Tanaka T, Nagata K, Takeyama M, Asuta M, Nakabayashi N and Masuhara E. 4-Meta opaque resin. A new resin strongly adhesive to nickel chromium alloy. *J Dent Res* 1981; **60**: 1697-1706.

Vallittu P K, Lassila V P (1992). Reinforcing of acrylic resin denture base material with metal or fibre strengtheners. *J Oral Rehabil* 1992; **19**: 225-230.

complete dentures

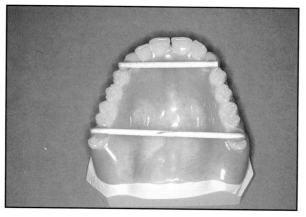

Fig. 3.4 The two parts of a fractured complete upper denture accurately relocated and secured in position with sticky wax and wooden stick reinforcement, ready for laboratory repair.

When the plaster has set, the wooden sticks and sticky wax are removed and the denture carefully lifted from the plaster model.

The plaster model is coated with a separating medium, for example sodium alginate. The fractured edges of the denture are bevelled and pumiced and then wetted with methyl methacrylate monomer to improve the bond strength of the repaired acrylic resin. Autopolymerised acrylic resin is normally used because of its easier processing technique however it is weaker than heat polymerised acrylic resin. The denture is reassembled on the plaster model and autopolymerised acrylic resin is applied to the fractured edges in slight excess. The model is placed in a hydroflask to effect an optimal cure. Once processed excess resin is removed and the denture polished.

If the denture is fractured into multiple fragments, it may be necessary to reposition the larger of the fragments intraorally and to take an in situ overall impression in alginate. If it is not possible to locate and accurately reassemble the fractured segments, repair should not be attempted and a new denture made. For the repair of fractured or missing teeth, an impression of the opposing dentition and/or denture is required to ascertain the correct occlusal rela-

complete dentures

Aesthetic denture problems

Facial appearance has a profound effect on many facets of human life. The features most commonly associated with facial attraction are the eyes and mouth. Good dental appearance is important to most individuals' psychological well being. Dental irregularities in children and adults may be treated as objects of fun which can, in extreme circumstances, lead to psychological disturbance. The loss of natural anterior teeth can be devastating for some individuals and their artificial replacement is therefore essential.

The aim of this chapter is to discuss the choice of anterior teeth for complete dentures and to assist the general dental practitioner in achieving satisfactory aesthetic results. The importance of correct anterior tooth position is emphasised.

Choice of anterior teeth

One of the main objectives in selecting and arranging artificial teeth is to produce a prosthesis which defies detection. With an appropriate degree of clinical expertise it should be a relatively straightforward procedure to select artificial teeth which blend harmoniously with the natural dentition. The choice of artificial teeth is, however, more complex when there are no natural teeth remaining and no pre-extraction records are available. The resultant prosthesis can be disappointing for both the dentist and the denture wearer, especially where there are conflicting ideals and expectations. Various factors have been suggested as indicators when selecting artificial teeth, for example:

- Simple observation which includes the temperamental theory whereby teeth were selected according to the individual's basal class, using the individual's body size, body form, colour of eyes and hair, disposition and character. In general, masculinity is characterised by vigour, boldness and hardness and femininity in terms of roundness, softness and smoothness. Dentogenics, a theory with its origins in the temperamental theory, derived from observation based on the sex, personality and age of the patient to produce highly personalised prostheses; Williams' classification, in which it was suggested that there was a correlation between face form and tooth form. Williams was convinced that this natural relationship was the key factor in determining desirable anterior tooth form. Most manufacturers' artificial tooth mould guides are divided into square, tapering and ovoid forms according to this classification.
- The use of measuring devices to determine face and tooth size/shape making selection quick and easy. These devices were generally placed on the patients head and the width/length of the face measured using bars inscribed with a scale and a pointer. From these readings teeth of suitable size and form could be selected.
- Patient surveys whereby individual preferences have been used as guides in the selection of anterior teeth and their arrangement. Surveys of individuals' preferred choices in the size, shape and arrangement of anterior teeth have revealed significant differences in perception when compared to professional concepts and ideals.
- The use of sectional casts with ideal compositions of the natural maxillary anterior teeth has been suggested as a suitable guide when determining anterior tooth arrangement. A number of mould guides are placed intra-orally until a satisfactory appearance is perceived.
- Combination theories encompassing the 'best' features of previously established methods have been used and reported to result in a higher success rate in achieving dentofacial harmony.
- Computer technology and formulae have been utilised in an attempt to determine precisely the degree of correlation between factors. Typically, these methods incorporated taking readings from anatomic landmarks on the teeth/face, digitising these shapes and using computer programs to derive shape plots. Conformity is determined by comparing width ratios between the two factors.
- Examination of previous dentures coupled with the copy denture technique.
- Use of old photographs which show the patient smiling with natural teeth.

There is, however, no universally accepted aesthetic factor that can be reliably used for the selection of artificial teeth. Leon Williams classification of facial and anterior tooth forms is one of the most significant contributions to denture aesthetics. Arguably it remains the most universally accepted method of determining anterior tooth form for edentulous and partially edentulous patients.

> **The aim of this chapter is to discuss the choice of anterior teeth for complete dentures and to assist the general dental practitioner in achieving satisfactory aesthetic results**

complete dentures

When choosing the size of the anterior teeth it has been suggested that the overall width of the central incisors is similar to the width of the philtrum of the upper lip. An extension of a line drawn from the inner canthus of the eye to the alar of the nose should pass through the upper canine. It is helpful to consider the skeletal build of the individual when choosing an appropriate size of tooth, for example, a larger skeletal build would be more suited to a larger tooth than one which is smaller and more artificial in appearance.

Colour of anterior teeth

The natural dentition is not monochromatic within the oral cavity but by contrast is polychromatic. There is a graduation of colour and of shades, significantly darkening with age. Within different age groups variations in lightness or darkness of complexion require corresponding adjustments to the selected colour of teeth. The colour of tooth substance is basically yellow due the underlying dentine showing through the translucent enamel. Within any one colour it is possible to produce lighter and darker shades. The addition of red makes this basic shade warmer and the addition of blue makes it cooler. The appropriate colour of artificial tooth should be chosen from a shade guide viewed in natural light. The artificial tooth should be moistened and held just inside the patient's mouth. The choice of shade should take into account the patient's comments however the dentist must rely on his clinical expertise and offer guidance as to the most suitable colour. A previous satisfactory denture can be of help.

Position of anterior teeth

It is very common to see major faults in the positioning of upper anterior teeth on complete dentures. This together with underextension of the bases is probably the most obvious error in denture construction. It is often a failure to comprehend the basic concepts of anatomy relating to oral aesthetics that is at the root of the problem. When the complex of muscles of facial expression, which intertwine at the corners of the mouth, is not held in proper physiological position by the anterior teeth and base material, the face droops. The resultant facial expression is typical of the 'denture look' so common in those who wear inadequate complete dentures. The corners of the mouth are turned down, the vermillion border of the lips is diminished, the nasial labial folds sag and wrinkles develop in the skin above the upper lip (fig. 4.1).

A basic understanding of the correct anterior tooth position is essential for good aesthetics. Correct positioning of the teeth is not time consuming and is well worth the effort. In the natural dentition upper anterior teeth, in a normal class I dental base relation, lie at an angle of 105° to the Frankfurt Plane and the lower anterior teeth are at an angle of 90° to the lower border of the mandible. It is the labial surfaces of the upper anterior teeth that should normally provide support for the upper lip not the flange.

An important landmark in the positioning of the anterior teeth is the incisive papilla, a structure which is affected very little by resorbtion and remains in a reasonably constant position (fig. 4.2). To reproduce the original position of the natural teeth the incisal edge of the upper central incisor should lie approximately 8-10 mm anterior to the centre of this landmark in a normal class I incisor relationship. If the anterior teeth are in the correct position, a line drawn between the middle of the canines should lie close to the middle of the incisive papilla (fig. 4.3). Support for the lips is largely provided by the teeth and attempts to plump out the buccal flange of an upper denture will not compensate for, nor disguise insufficient

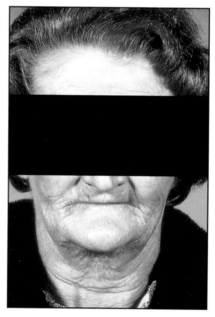

Fig. 4.1 The inadequate complete denture in this individual does not provide appropriate support for the lips and cheeks. There is sagging and wrinkling of the tissues around the lips which has resulted in a 'denture look'.

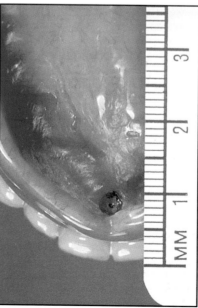

Fig. 4.2 The incisive papilla is an important landmark in the positioning of the anterior teeth. In this case it was possible to place the upper central incisors in the recommended position of 8-10 mm anterior to the landmark. It is not always possible to use this recommended relationship as a compromise with function may dictate a more palatal position for the anterior teeth.

support by anterior teeth. If the anterior teeth are placed too far palatally the upper lip may fall in and there is a possibility of folding at the corners of the mouth and, when there is an associated denture induced stomatitis, the subsequent development of angular stomatitis (chelitis). This is probably the most common cause of erythema and breakdown at the corners of the mouth and not overclosure as stated in many standard prosthetic textbooks.

The lips should meet at the line junction between transitional and oral mucosa and the full breadth of the red margins can then be seen. On either side of the philtrum the tissues fall away but more abruptly so at the upper end due to a depression at the alae of the nose which corresponds to the area marking the root of the lateral incisor. In the average individual the angle between the columella of the nose and the lip is approximately 90° although this may vary depending on the Angles classification. For example, in class II Division II malocclusion where the upper anterior teeth are retroclined, the nasiolabial angle may be more than 90°.

In the upper arch, the central incisors are a key point in determining the overall aesthetics as they are a base line from which other teeth are set. The long axis of the central incisors should be almost vertical when viewed from the front and inclined downwards and slightly forwards when viewed from the side. The lateral incisor is inclined with its neck positioned more distally than the incisal edge. The canines are the corner stone of the arch and are arranged vertically when viewed in all planes. It is the neck of the canine that gives support to the corners of the mouth. In the lower arch the central and lateral incisors are inclined slightly forwards and towards the midline and the canines slightly towards the midline with the necks prominent.

When constructing a complete denture the upper occlusal rim should be adjusted labially to provide adequate lip support and indicate the correct position of the anterior teeth for a good aesthetic result. Failure to do so may lead to a thinning of the vermillion border, a convexity of the philtrum and fullness in the alae areas of the nose. The height of the upper occlusal rim should be adjusted to the desired length of the teeth, paying particular attention to the upper lip length and the amount of tooth the patient wishes to show. Textbook suggestions that the average visibilities of the central incisors in relation to the upper lip line (at rest) are: young person: 2 mm below lip; middle age: 1-1.5 mm below lip; old age: 0-2 mm above lip are no more than guidelines and should not be followed rigidly. The lower rim should be adjusted to the correct occlusal vertical dimension (OVD). To the experienced dentist the correct OVD 'looks right' whereby the mentalis muscle is not straining to produce a lip seal as a result of an increase in height and there is not excessive folding at the corners of the mouth indicative of a reduction.

Speech can also be used to assess the vertical dimension and tooth position. The sound 's' will be unsatisfactory if the teeth touch when the sound is produced. Difficulties in pronouncing 's' very often occur due to an excessive overjet and this can usually be corrected by repositioning the upper anterior teeth more palatally or the lower anterior teeth further labially or by a combination of both. The individual will also have difficulty in pronouncing the fricatives 'f' and 'v' if the upper anterior tooth position is incorrect, for example if the upper anterior teeth are too far back or the incisor tips too low.

Choice of tooth material

In order to produce natural looking artificial teeth, the colour, optical properties and distribution of enamel and dentine must be reproduced. Enamel is a beautifully translucent material and no artificial material as yet can match its optical properties. However, porcelain and acrylic resin teeth can give satisfactory results if handled properly and few would dispute the superb aesthetics that can be achieved with porcelain. Despite this porcelain is not without its limitations: the teeth are brittle and prone to fracture; they have to be mechanically attached to the denture base, can be noisy compared to acrylic resin and are relatively expensive. They are of course much more difficult to adjust when setting in the laboratory and require absolute precision when establishing the occlusion in the clinic. In recent years acrylic resin teeth have been improved both in terms of mechanical properties, in particular abrasion resistance, and aesthetics. These factors, together with cost, may explain the routine use of acrylic resin teeth in complete denture construction.

Personalisation of the appearance

For the individual who is keen to achieve a natural appearance there is much that can be done to reproduce features present in their natural

Further reading

Bell R A. The geometric theory of selection of artificial teeth: Is it valid? *JADA* 1978; **97**: 637-640.

Brisman A S. Esthetics. A comparison of dentists' and patients' concepts. *JADA* 1980; **100**: 345-352.

Brodbelt R H W, Walker G F, Nelson D, Seluk L W. Comparison of face shape with tooth form. *J Prosthet Dent* 1984; **52**: 588-592.

Esposito S J. Esthetics for denture patients. *J Prosthet Dent* 1980; **44**: 608-615.

French F A. The selection and arrangement of the anterior teeth in prosthetic dentures. *J Prosthet Dent* 1951; **1**: 587-593.

Frush J P, Fisher R D. How dentogenic restorations interpret the sex factor. *J Prosthet Dent* 1956; **6**: 160-172.

Frush J P, Fisher R D. How dentogenic restorations interprets the personality factor. *J Prosthet Dent* 1956; **6**: 441-449.

Frush J P, Fisher R D. The age factor in dentogenics. *J Prosthet Dent* 1957; **7**: 5-13.

Hughes G A. Facial types and tooth arrangement. *J Prosthet Dent* 1951; **1**: 82-95.

Krajicek D D. Guides for natural facial appearance as related to complete denture construction. *J Prosthet Dent* 1969; **21**: 654-662.

Marunick M T, Chamberlain B B, Robinson C A. Denture aesthetics: an evaluation of laymen's preferences. *J Oral Rehabil* 1983; **10**: 399-406.

Mavroskoufis F S, Ritchie G M. The face-form as a guide for the selection of maxillary central incisors. *J Prosthet Dent* 1980; **43**: 501-505.

Sellen P N, Jagger D C, Harrison A. The correlation between selected factors which influence dental aesthetics. *J Primary Dent Care* 1998; **5**: 55-60.

Seluk L W, Brodbelt R H W, Walker G F. A biometric comparison of face shape with denture tooth form. *J Oral Rehabil* 1987; **14**: 139-145.

Vallitu P K, Vallitu A S J, Lassila V P. Dental aesthetics — a survey of attitudes in different groups of patients. *J Dent* 1996; **24**: 335-338.

Vig R G. The denture look. *J Prosthet Dent* 1961; **11**: 9-16.

Williams J L. The temperamental selection of artificial teeth, a fallacy. *The Dental Digest* 1914; **20**: 63-75, 125-134, 185-195, 243-259, 305-321.

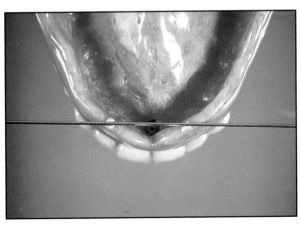

Fig. 4.3 A line connecting the tips of the canines should lie close to the middle of the incisive papilla if the anterior teeth are in the correct position.

complete dentures

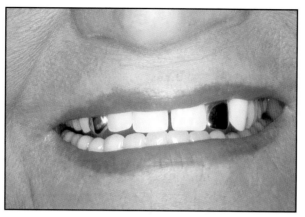

Fig. 4.4 The aesthetics of an upper complete denture can be modified to suit the aesthetic needs of an individual. In this case gold restorations have been incorporated.

Fig. 4.5 Selection of an appropriately pigmented denture base material for a particular racial group will enhance the aesthetics.

dentition, for example the introduction of a diastema, creating irregularities with tilting and overlapping of teeth, selective incisal edge grinding to simulate tooth wear, the incorporation of restorations (fig. 4.4), the use of stains and the simulation of gingival recession where appropriate. Different coloured denture base resins are available to reflect the variation in colouration in different racial groups (fig. 4.5).

Most individuals do not want others to realise that they are wearing complete dentures. An understanding of the principles of aesthetics coupled with an appropriate degree of clinical expertise should enable the dentist to make natural looking complete dentures.

complete dentures

5 The dirty denture

It is not surprising that denture wearers are confused by the vast array of denture cleansers available for purchase over the counter. Should they brush their dentures with powders or pastes or perhaps soak them in a solution? Which method is best? Each product comes with claims for its relative efficacy — making the choice difficult. Surveys show that many denture wearers experience difficulty in cleaning their dentures satisfactorily and some continue to wear dirty dentures (fig. 5.1).

Individuals may attend for replacement of their dentures due to general deterioration of the denture base material, often as a result of the misuse or abuse of approved cleaning methods or the use of alternative regimes such as frequent soaking in household bleach (fig. 5.2) or the use of a harsh brush and/or abrasive paste (fig. 5.3). Possible explanations for these findings are either that the dental profession is not giving correct advice on approved cleaning methods or that the advice given is not being followed. A recent survey reported that of 100 individuals wearing complete dentures, partial dentures or a combination of the two, 46% claimed that they had never received advice on how to clean their dentures. Forty-six percent had received advice on denture cleaning from their general dental practitioner (GDP) and the remainder from a relative (4%), hygienist (2%) and a pharmacist (2%).

The aim of this chapter is to provide an update on the types of denture cleansers available for purchase over the counter, and to make recommendations for their use such that the GDP can give informed advice to the denture wearer.

Requirements of a denture cleanser
Ideally a denture cleanser should:
- be non toxic and non irritant
- be easy to apply and easy to remove without residue
- remove the organic portion of denture deposits (15-30% of the total deposit on the denture consisting mainly of mucoproteins)
- remove the inorganic portion of denture deposits (70-85% of the total deposit consisting mainly of calcium phosphate, calcium carbonate and lesser quantities of other phosphates)
- be harmless to all the materials used in denture construction and maintenance eg poly (methyl methacrylate), cobalt chromium, stainless steel, temporary soft materials and permanent resilient soft lining materials
- be stable on storage with a long shelf life
- be bactericidal, fungicidal and viricidal
- be relatively inexpensive.

Types of denture cleansers
Denture cleansers can be classified according to their mode of action, that is mechanical or chemical.

Mechanical action
Included in this category are the abrasive pastes, used together with brushes, and the ultrasonic cleaners. The latter could also be grouped with the chemical cleansers depending on the solution used in the ultrasonic bath. Examples of the cleaners with a mechanical action are:
- Dentucreme (Stafford Miller Ltd, Hatfield, Herts, UK).
- Boots Denture Toothpaste Original (The Boots Company PLC, Nottingham, UK).

The aim of this chapter is to provide an update on the types of denture cleansers available for purchase over the counter, and to make recommendations for their use such that the GDP can give informed advice to the denture wearer.

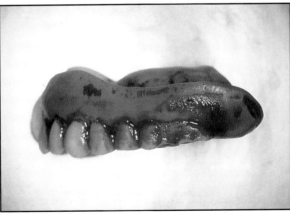

Fig. 5.1 Plaque and debris on a denture disclosed with erythrosin dye.

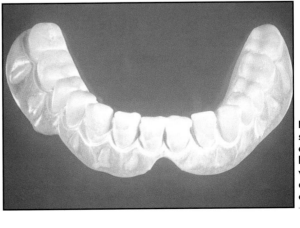

Fig. 5.2 Frequent soaking of the denture in household bleach has caused whitening and deterioration of the acrylic resin.

complete dentures

- Boots Denture Toothpaste Minty.

These cleaners are easy to use, relatively effective if used with an appropriate degree of expertise and not too expensive. However over enthusiastic brushing techniques coupled with harsh brushes can lead to severe damage of the denture base material. Another disadvantage is their unsuitability for use by those individuals with impaired manual dexterity, such as the handicapped, where an ultrasonic cleaner or a chemical cleanser is a more appropriate choice.

Chemical action

This category can be subdivided into five groups depending on the constituents and mechanism of action:
1. Effervescent Peroxides
2. Alkaline Hypochlorites
3. Acids
4. Disinfectants
5. Enzymes.

1. Effervescent peroxides. Examples are:
- Steradent Original, Steradent Minty, Steradent Deep Clean Tablets (Reckitt Dental Care, Reckitt and Colman, Hull, UK).
- Boots Effervescent Original, Boots Double Action Tablets (The Boots Company PLC, Nottingham, UK).
- Superdrug Original, Superdrug Minty, Superdrug Extra Strength Tablets (Superdrug Stores PLC, Croydon, Surrey, UK).
- Super Efferdent Tablets (Warner Lambert HealthCare, Eastleigh, Hampshire, UK).
- Boots Denture Cleansing Powder (The Boots Company PLC, Nottingham, UK).
- Steradent Denture Cleansing Powder (Reckitt Dental Care, Reckitt and Colman, Hull, UK).

These are supplied in two main forms, powder or tablets, and are used by mixing with water. They can be alkaline, acidic or neutral although the manufacturers tend not to specify which type on the packaging. These cleansers are rapid in action, simple to use and relatively effective if used correctly and regularly on dentures which do not already have heavy stains or calculus deposits. Problems can arise if very hot water is used with the denture cleanser. This coupled with prolonged periods of soaking can lead to excessive bleaching of the poly (methyl methacrylate) denture base material. The effervescent peroxides are a popular choice among denture wearers. They have a possible additional mechanical cleansing action due to the bubbles created by the release of oxygen which may dislodge debris but there is limited antibacterial effect and they will not remove calculus. Some products incorporate enzymes (proteolytic and yeastlytic) which may increase their effectiveness by helping the degradation of proteins in the plaque matrix.

2. Alkaline hypochlorites. Examples are:
- Dentural, (Martindale Pharmaceuticals, Romford, Essex, UK).
- Milton (Procter and Gamble Ltd, Egham, Surrey, UK; Not marketed as a denture cleanser but in common use).

Alkaline hypochlorite can be recommended for its superior cleansing properties. It is effective in the dissolution of plaque and can have inhibitory effects on calculus formation because of its effect on the plaque matrix. It has superior stain removal properties due to its bleaching action and it has some bactericidal and fungicidal properties. Disadvantages of the hypochlorites are possible excessive bleaching of acrylic resin and corrosion of metal with prolonged soaking together with the residual taste and odour after use.

3. Acids. Examples are:
- Denclen (Procter and Gamble Ltd, Egham, Surrey, UK).
- Deepclean (Reckitt Dental Care, Reckitt and Colman, Hull, UK)

These are supplied as a liquid in a plastic container which has a brush attachment. They are less popular but are useful for stubborn stains and calcified deposits. They can cause corrosion of metallic components of dentures.

4. Disinfectants. An example is:
- Chlorhexidine (Smithkline Beecham Consumer Healthcare, Brentford, UK)

This cleanser is recommended as an adjunct in the treatment of denture induced stomatitis. It is suggested that the dentures are soaked in chlorhexidine solution for 15 minutes twice daily. Prolonged use, however will lead to brown staining due to dietary chromogens (tea, coffee, red wine etc).

5. Enzymes. An example is:
- Polident (Kobayashi Block Co Ltd, Osaka, Japan).

The incorporated enzymes (proteolytic and yeastlytic) may increase the effectiveness of a cleanser by helping the degradation of proteins in the plaque matrix.

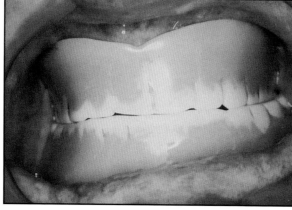

Fig. 5.3 Overenthusiastic cleaning by brushing this denture with an abrasive paste has worn away the surface detail and the definition between the teeth and denture base.

complete dentures

A recommended approach to denture hygiene

It is recommended that all types of dentures are rinsed with water after every meal and debris removed by brushing with a soft brush, soap and cold water.

Acrylic resin dentures

Acrylic resin dentures should be soaked in an alkaline hypochlorite solution for 20 minutes in the evening and should then be rinsed and soaked in cold water overnight. This cleaning regime may be coupled with the occasional use of an acid cleaner for stubborn stains or calcified deposits.

Metal based dentures

For metal based dentures eg cobalt-chromium, the alkaline peroxides can be a suitable cleanser. Alkaline hypochlorites may be used but care must be taken as prolonged soaking can cause discolouration of the metal as the solution is potentially corrosive. The acidic cleaners are contraindicated as these will cause corrosion of the metal component.

Temporary soft materials

Some denture wearers will use a temporary soft material [eg Viscogel (Dentsply Ltd, Weybridge, UK) or Coe Comfort (Coe Laboratories Inc, Chicago, USA)] in conjunction with their denture either as a tissue conditioner or as a temporary soft lining material. For optimum conditioning of the tissues the material should be soft and plastic, however the incorrect use of approved denture cleansers or the use of inappropriate cleaning regimes can cause rapid deterioration of these materials and they become rough, hard, discoloured and exhibit malodour. Brushing, with or without abrasive pastes, should be avoided as should the use of effervescent peroxides which may cause bubbling and deterioration of the surfaces (fig. 5.4). A recommended regime is to rinse the dentures and soft material after each meal and soak daily for 20 minutes in hypochlorite. The material gradually hardens due to the progressive leaching and evaporation of some of its components and therefore should be replaced frequently. There may be a residual odour and taste resulting from the long-term use of alkaline hypochlorites with temporary soft materials.

Permanent resilient soft lining materials

For the two categories of permanent soft lining materials, silicone (eg Molloplast B, Molloplast KG, Kostner and Co, Karlshue, West Germany) and acrylic resin (eg Coe Super Soft, Coe Laboratories Inc, Chicago, USA), similar cleaning procedures are recommended as for the temporary materials although they can be brushed lightly with a soft brush and soap and water. The other main types of denture cleansers available are not recommended as they appear to have a deleterious effect on the soft linings (fig. 5.5). If a metal strengthener is incorporated into the denture the period of soaking in alkaline hypochlorite should not exceed 10 minutes (according to the manufacturer's instructions) due the possibility of damage to the metal component.

Denture fixatives

Denture fixatives are a group of materials used to improve the retention of dentures. For individuals for whom retention is a great problem the regular use of fixatives becomes the norm. If used according to the manufacturers' instructions and applied to the denture with an appropriate degree of expertise, these materials can be a useful aid to denture retention. However the misuse of denture fixatives, such as the repeated addition of fixative, in the absence of good denture hygiene, can result in a denture with malodour and the accumulation of food debris, plaque and calculus with its harmful effects on the mucosa. Some fixatives will selectively support the growth of certain microorganisms whilst inhibiting the growth of others. It is important that all fixative is removed from the denture and the denture cleaned before fresh material is applied.

A recommended cleaning regime is to remove the denture before retiring to bed and to remove the denture fixative by brushing with a soft brush and warm water. The denture should them be cleaned by one of the approved

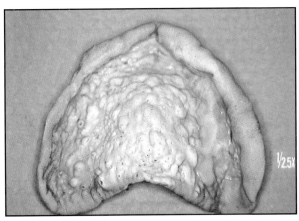

Fig. 5.4 Soaking in an effervescent peroxide denture cleanser has resulted in surface bubbling and deterioration of a temporary soft lining material (tissue conditioner).

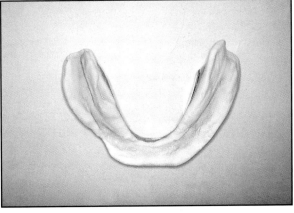

Fig. 5.5 An example of a permanent soft lining material which has become less resilient and bleached as a result of prolonged soaking in an alkaline hypochlorite denture cleaning solution.

complete dentures

Further reading

Davenport J C, Wilson H J, Spence D. The compatibility of soft lining materials and denture cleansers. *Br Dent J* 1986; **161:** 13-17.

Gates W D, Goldschmidt M, Kramer D. Microbial contamination in four commercially available denture adhesives. *J Prosthet Dent* 1994; **71:** 154-158.

Hoad-Reddick G, Grant A, Griffiths C. Investigation into the cleanliness of dentures in an elderly population. *J Prosthet Dent* 1990; **64:** 48-52.

Jagger D C, Harrison A. Denture cleansing — the best approach. *Br Dent J* 1995; **178:** 413-417.

Jagger D C, Harrison A. Denture fixatives — an update for general dental practice. *Br Dent J* 1996; **180:** 311-313.

Jagger D C, Harrison A. Complete dentures — the soft option: an update for general dental practice. *Br Dent J* 1997; **182:** 313-317.

Lamb D J. The effects of karaya gum on tooth enamel. *Br Dent J* 1980; **150:** 250-252.

Nakamoto K, Tamamoto M, Hamada T. Evaluation of denture cleansers with and without enzymes against *Candida albicans*. *J Prosthet Dent* 1991; **66:** 792-795.

Ritchie G M. A report of dental findings in a survey of geriatric patients. *J Dent* 1973; **1:** 106-118.

Stafford G D, Russell C. Efficiency of denture adhesives and their possible influence on oral micro-organisms. *J Dent Res* 1971; **50:** 832-836.

methods.

Denture cleanliness is essential to prevent malodour, poor aesthetics and the accumulation of plaque/calculus with its harmful effects on the mucosa. A recommended approach to denture hygiene should be adopted for denture cleanliness and to prevent damage to the denture base material that can be caused by the use of inappropriate cleaning regimes.

6 The 'problem' denture

complete dentures

The experienced general dental practitioner will be aware that some individuals can function well and are content with dentures that have errors which are obvious to the clinician. However others have great difficulty with, or are intolerant of, dentures which have been constructed with a high level of clinical and laboratory expertise. For example, the individual who attends for the provision of replacement dentures with a bag containing a number of sets of unsatisfactory complete dentures can be a challenge for even the most experienced practitioner (fig. 6.1). Unfortunately the provision of a technically 'perfect' denture may not solve the denture problem for that individual.

Psychiatric disorders are common and are often undiagnosed. It is inevitable that therefore some individuals attending for the provision of complete dentures will be suffering from, and presenting with, signs and symptoms of psychiatric illness. Sometimes individuals with psychiatric problems present with behaviour that is so abnormal that the presence of a disorder is obvious (for example, schizophrenia). However, in other cases the underlying condition may not be so apparent. On occasions problems relating to dentures may be the only symptoms of an underlying psychiatric condition.

A basic knowledge of psychiatric disorders is necessary for the recognition of presenting symptoms which may themselves be a symptom of an underlying psychiatric disorder; for the formulation of treatment plans with realistic, achievable goals and if appropriate for referral to specialist dental and/or mental health professionals.

The General Dental Council recognises the importance of psychological/psychiatric disorders and has introduced the topic as a mandatory part of the undergraduate curriculum. Further postgraduate opportunities and experience can be gained through vocational, general professional and specialist training. The aim of this chapter is to provide an outline of psychiatric disorders of relevance to the general dental practitioner and to discuss some important examples of oral manifestations of psychiatric disease relevant to prosthetic dentistry.

Classification of Psychological Disorders

Psychiatric disorders can be classified according to The Diagnostic and Statistical Manual of Mental Disorders, 3rd Revision, DSM-III-R of the American Psychiatric Association (1987) and the International Classification of Diseases ICD10 (World Health Organisation, 1993). Psychological disorders can broadly be classified into four categories:
- Neuroses
- Personality disorders
- Psychoses
- Others.

Neuroses

Neuroses are minor mental disorders in which contact with reality and insight are retained. Anxiety is an unpleasant feeling of fearful anticipation. It is a universal emotion that is a useful defence mechanism. If this so-called normal anxiety becomes excessive it can have a negative or ill effect on a person and his or her behaviour. Such an anxiety is termed morbid anxiety and underlies all neurotic conditions.

Neuroses can be subdivided into different types, for example; generalised anxiety, depressive, obsessional, hypochondriacal, hysterical, and phobic neuroses. Obsessional anxiety neurosis is a morbid anxiety which cannot be controlled, where there is a feeling of compulsion and preoccupation with thoughts and acts and obsessive traits, such as repeated senseless behaviour. Individuals have insight and realise that their behaviour is illogical and may try and conceal it. Depressive neurosis is a mild form of depressive illness and is characterised by pessimism, low self esteem, sleep disturbance and loss of libido. Hysterical neurosis is where the individual's physical complaints have no demonstrable organic basis, and hypochondriacal neurosis is characterised by a morbid pre-

The aim of this chapter is to provide an outline of psychiatric disorders of relevance to the general dental practitioner and to discuss some important examples of oral manifestations of psychiatric disease relevant to prosthetic dentistry.

Fig. 6.1 The unwelcome sight of a bag of unsuccessful complete dentures which have been provided by a number of dental practitioners.

occupation with disease and physical complaints. Phobias are neuroses characterised by fears focused on a particular object or situation, which are out of all proportions to the demands of the situation and which do not respond to reason.

Personality disorders
Personality disorders are chronic peculiarities of character or maladaptation to life where personality traits take over and form a large feature of the personality. There is no loss of insight and individuals are in touch with reality. Examples are obsessive-compulsive characterised by high standards, rigidity, chronic worry and precision; hysterical with immaturity and attention seeking; schizoid characterised by shyness, secrecy, isolation and lack of friends; paranoid with a fear of persecution and suspicion and psychopathic also referred to as antisocial behaviour, characterised by selfishness, disloyalty and conflict.

Psychoses
Psychoses refer to more severe, far reaching and prolonged forms of mental disorders where the individual is divorced from reality and lacks insight. These are subdivided into functional psychoses (eg mood, schizophrenia and allied disorders) and organic psychoses.

Within the organic psychoses, the acute organic states have impairment of orientation, grasp of knowledge and concentration. The chronic organic states include the dementias, which are characterised by global disruption of personality with the gradual development of abnormal behaviour, loss of intellect, mood changes, blunting of emotions and failure to learn. Eventually there is a reduction in self care, restless wandering, paranoia and incontinence.

The mood psychoses include severe (psychotic) depression, mania, manic depression and schizophrenia or allied disorders. Severe depression is different from neurotic depression and is characterised by early morning wakening, loss of appetite, loss of libido, pessimism, delusions or amplification of pain together with obsessional symptoms, phobias, poor memory and hypochondriasis. In comparison, mania is an elevation of mood, irritability, rapid speech, and increase in mentation. Manic depressive psychoses exhibit both mania and depression. Schizophrenia presents with disturbances of thought, mood and conduct. Hallucinations occur and delusions are common, especially delusions of persecution (paranoia). Individuals show withdrawal from life, shyness, suspicion and have bizarre ideas and theories. They may become alcoholic and commit suicide.

Others
The remaining group are largely disorders of behaviour, for example alcohol abuse and drug abuse. Psychiatric complications can occur in association with alcohol and / or drug abuse, for example; alcoholic dementia, alcoholic hallucinosis and mood disorders, such as depression and anxiety.

Orofacial manifestations of psychiatric disease relevant to the construction of complete dentures
Frequently orofacial symptoms are a predominant manifestation of psychiatric disorders. For comprehensive coverage the reader is referred to the further reading list but some important examples are as follows:

Neuroses
Anxiety. Anxiety can produce muscle tension and pain. An underlying anxiety can be an important factor in temporomandibular joint dysfunction and bruxism can lead to complaints of soreness and pain under complete dentures because of excessive pressure put on the underlying mucosa and ridge.

Denture phobias. A phobia is a fear which is out of all proportions to the demands of the situation and which does not respond to reason. Phobias can be specifically focused on dentures, presenting as a fear of other people knowing that the individual wears dentures or as a fear of objects, such as dentures, in the mouth causing airway obstruction. The latter can be associated with an increased gag reflex. The majority of denture phobics can be treated successfully in general dental practice although handling them requires sympathy, tact and skill which will come with the appreciation of these disorders and experience. The use of behaviour therapy with gradual exposure to the stimulus is a useful approach. In cases of severe phobia it is appropriate to refer the individual for specialist help — preferably to a unit which provides both dental care and liaison psychiatric support.

Psychoses
Underlying depression. Disorders of taste and salivation are frequently associated with underlying depression. Depression can produce poor motivation, lack of interest and problems may be focused on the dentures. Severe depressive symptoms include hypochondriasis, delusions and amplification of pain. Xerostomia is a common side effect of antidepressant therapy and poor denture retention may be a complication. Psychotic states, especially schizophrenia, can be associated with delusional pain and bizarre symptoms. Psychotic disorders need treatment as early as possible.

Monosymptomatic hypochondriacal psychosis. This group of disorders is closely related to schizophrenia. Within this category are those individuals who present with 'phantom bite syndrome' where there is a delusion that the occlusion is abnormal. These individuals may present for complete dentures with a history of occlusal adjustments and extractions in an attempt to rectify the occlusion. Other examples are the delusion of oral infestation by insects or worms (Ekbom's syndrome) and 'body dysmorphic disorder' characterised by a subjective feeling of cosmetic defect in an individual of apparently normal appearance, which can be related to the perceived appearance of any part of the body, and can include the face and mouth. Individuals with body dysmorphic disorder can make excessive demands on the dentist to correct the shape, size and position of the natural or artificial teeth in their search for (unachievable) perfection.

Personality disorders. The difficult behaviour associated with personality disorders can make dental treatment problematic, for example the high standards, rigidity of approach and attention to minor detail expressed by individuals with obsessive compulsive disorders and the aggression and awkwardness of those individuals with passive/aggressive disorders. Handling these individuals requires tact, skill, patience and tolerance as well as avoiding confrontation and conflict.

Although a degree of flexibility in patient management will be productive the general dental practitioner must avoid being manipulated by the patients and being drawn into treatment methods he or she thinks unwise or unnecessary. Referral or discharge is often more appropriate in these cases.

Clinical management

With an appropriate degree of clinical expertise and sufficient insight into the underlying psychiatric problems, many individuals can be treated satisfactorily in general dental practice. Individuals who present with behaviour that is so abnormal that the presence of an underlying psychiatric disorder is obvious or who suffer from a mild form of mental illness that compromises their successful treatment in general dental practice, should, in association with their general medical practitioner, be referred for specialist care. A closer liaison between dental and mental health professionals and the early recognition that psychiatric treatment may be necessary as an adjunct to the provision of complete dentures, in some cases, may in the long term reduce the overall clinical treatment time.

Further reading

Chamberlain B B, Chamberlain K R. Depression: a psychologic consideration in complete denture prosthodontics. *J Prosthet Dent* 1985; **53:** 673-675.

Enoch M D, Trethowan W. *Uncommon psychiatric syndromes.* 3rd edn. Oxford, UK: Butterworth-Heinemann, 1992.

Enoch M D, Jagger R G. *Psychiatric disorders in dental practice.* Oxford, UK: Butterworth-Heinemann, 1994.

Kotwal K R. Beyond classification of behaviour types. *J Prosthet Dent* 1984; **52:** 874-876.

Scully C, Cawson R A. *Medical problems in dentistry.* Chapter 14. Psychiatric Disease. pp 379-405. Oxford, UK: Wright, 1987.

Appendix 1: increased gag reflex

The gag reflex is a protective response that prevents foreign objects or noxious material from entering the pharnyx, larynx or trachea. A heightened gag reflex (retching) can lead to difficulty in carrying out dental procedures. The severity of the problem can vary from mild retching, for instance, when taking dental impressions or wearing complete dentures, retching precipitated by external stimuli such as the sight of impression material being mixed or the sound of another individual retching, to severe uncontrolled retching. It is often a multifactorial problem and may have both a physical and/or psychological basis.

Some common causes of retching are:
- overextension of the posterior palatal border of an upper complete denture causing stimulation of the sensitive soft palate;
- poor retention and looseness of an upper complete denture (Chapter 1) resulting in movement of the base and stimulation of the gag reflex;
- excessive thickening of the posterior palatal border and tuberosity regions irritating the posterior aspect of the tongue;
- lingual encroachment of the lower molar teeth causing insufficient tongue space;
- external stimuli;
- systemic factors such as gastrointestinal disease;
- association with excessive intake of alcohol;
- psychological factors, for example phobias related to fear of objects in the mouth and airway obstruction.

Management

The management of retching must deal with both the physical and psychological causes. The majority of individuals can be treated satisfactorily in general dental practice. However, for a minority, retching may be so severe as to compromise the overall success of the dental treatment and these individuals should be referred to appropriate dental specialists working in association, if necessary, with a general medical practitioner and mental health professional.

Where the primary cause of retching is a fault associated with the denture, adjustments and correction of the fault(s) will usually solve the problem, (for example for a loose denture, correction of under or overextension or inappropriate tooth position). If the individual experiences difficulty in having impressions taken, a sympathetic and understanding approach by the dentist together with a carefully controlled impression technique is helpful. Time should be allocated to allow for discussion, reassurance and acclimatisation before the impression is taken.

When taking an impression, the dental chair should be adjusted so that the patient is in a comfortable upright position. Provided there are minimal undercut areas a satisfactory primary impression can usually be obtained using impression compound in a suitable stock impression tray. This material can be moulded quickly and removed after a short time, it also has a more acceptable texture than alginate. Since the lower is often better tolerated it is usually advisable to complete this first to gain the patient's confidence. For the special tray and working impression, a close fitting tray (as opposed to a spaced tray) should be used when ever possible, as it is smaller and better tolerated by the patient.

The special tray should be correctly trimmed to allow for appropriate border moulding with greenstick tracing compound. Zinc oxide eugenol is the impression material of choice, however a light bodied silicone rubber may be used. The tray should not be overloaded as excess will be expressed from the tray and may irritate the palatal tissues and precipitate retching. Once the tray has been inserted into the mouth and seated, it should be held firmly in position. The patient should be reassured, in a calm confident manner, to breathe deeply through the nose and not the mouth. If possible the tray should not be removed until the material has set (it will only have to be repeated!). A kidney bowl (in case of retching and subsequent vomiting) should be kept out of sight as this may itself precipitate retching. If a spaced special tray is required alginate is the material of choice. The setting time may be modified by using warm water to ensure a faster set.

For individuals where retching is a major problem the use of supportive sedation techniques (inhalation or intravenous) has been recommended. However, although the use of these techniques may overcome the initial problem of obtaining a satisfactory working impression, they do not solve the problem of adjusting to and being able to wear complete dentures. The use of hypnosis can be a useful adjunct to the treatment of retching. Also relax-

complete dentures

Further reading

Barsby M J. The use of hypnosis in the management of gagging and intolerance to dentures. *Br Dent J* 1994; **176:** 97-102.

Freidman M H, Weintraub M I. Temporary elimination of gag reflex for dental procedures. *J Prosthet Dent* 1995; **73:** 319

Hoad-Reddick G. Gagging — a chairside approach to control. *Br Dent J* 1986; **161:** 174-176.

Muir J D, Calvert E J. Vomiting during the taking of dental impressions. Two case reports of psychological techniques. *Br Dent J* 1988; **165:** 139-141.

Wright S M. An examination of factors associated with retching in dental patients. *J Dent* 1980; **7:** 194-207.

ation therapy, anxiety control, controlled breathing, conditioning, desensitisation and confidence pattern control have been described in the dental literature.

When the provision of complete dentures has failed, a training plate (or plates) is a treatment option to allow the patient to gradually adapt to wearing a prosthesis. Initially the training plate will consist of a simple heat cured acrylic resin base plate constructed on the working model. Teeth are progressively added to the base plate over a period of time as the patient adapts to wearing the plate and gradually a template for a complete denture is formed. This template(s) can then be used to construct dentures using a copy denture technique (see Appendix 4).

The use of a horseshoe design for the upper denture may be an option for those individuals with good alveolar ridges with the potential for reasonable retention and stability who cannot tolerate full palatal coverage. However this type of denture is often less retentive than full coverage and inherently weaker and prone to fracture. Unfortunately, regardless of the time and effort spent, a few individuals will never be able to tolerate wearing dentures and this has to be accepted.

Position of the post-dam

The post-dam should be positioned at the junction of the hard and soft palate, a position referred to as the vibrating line. This line is the boundary between the vibrating and non-vibrating part of the palatal mucosa. The narrow strip of mucosa at this boundary is immobile. The junction of the hard and soft palates is visible intraorally as a faint line and can be determined by asking the patient to say 'aah'. The position of a post-dam on an exisiting denture may be clearly visible intraorally as an indentation in the palatal mucosa but should only be used as a guide and should not be copied. The displaceability of the palatal tissues can be determined by simple digital palpation or with the aid of a ball-ended instrument.

The position of the vibrating line should be marked on the working model in the form of a cupid's bow (Fig A1.1). A groove for the post dam can be carved with an Ash 5 wax carving instrument. The depth of the groove will be dependent on the mucocompressibility of the tissues. The post dam should be deeper in the areas of greater mucocompressibility and shallow where the mucosa is tightly bound. Laterally it should extend distal to the maxillary tuberosities into the hamular notches for maximum retention.

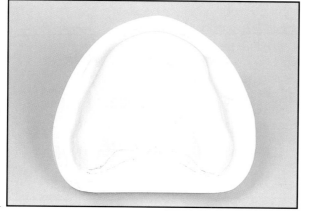

Fig. A1.1 An outline of the post dam is carved on the working model. It is important that the posterior border of the finished denture does not extend onto the moveable soft palate which can precipitate retching and that the denture base is kept as thin as possible in this region.

complete dentures

Appendix 2: the neutral zone

A complete denture may be displaced if it interferes with the surrounding oral musculature. It will be most stable if it lies within the neutral zone, a zone of minimal conflict between the tongue on one side and the lips and cheeks on the other. Although the zone influences both upper and lower dentures, because of the greater area for support in the maxilla and the greater potential for interference with the musculature in the lower jaw, it is usually the lower complete denture which is most easily displaced. The technique is therefore almost always applied to the lower denture.

The position of the neutral zone can vary with post extraction changes, for example; lateral spreading of the tongue following extraction of the lower posterior teeth will force the neutral zone buccally, and may well also reduce its buccolingual width, and extensive mandibular anterior bone resorbtion resulting in prominence of the mentalis muscle fibres, will have a local effect and move the zone lingually. The clinical and laboratory technique is costly in time and resources compared to the conventional approach and it is therefore generally reserved for those cases which have either already failed because of design faults or those that the dentist predicts are unlikely to be successful unless the technique is used to overcome the problems of a conventional approach with its inherent functional difficulties. It is possible to record the position of the neutral zone using a variety of techniques. One suitable technique for use in general dental practice, is described.

Clinical technique

Primary and working impressions are taken and centric jaw relationship recorded as for routine complete denture construction. An autopolymerising acrylic resin lower base plate with molar pillars is made on the working cast at the correct occlusal vertical dimension, dictated by the occlusal rims (fig. A2.1).

At the next clinical stage, tissue conditioner, for example, Viscogel (De Trey Division, Dentsply Ltd, Weybridge, UK) is applied to the baseplate and the patient instructed to swallow and talk, so that the functional position of the muscles is recorded. Additional powder will be required in the Viscogel mix so that it will support its own weight and allow an occlusal rim like form to be built up on the baseplate. Its consistency however will permit the tissues to mould and shape it during a range of normal functional muscular activity over an extended period. When the material has reached a firmer state it can be removed from the mouth and inspected. The neutral zone impression is carefully handled to avoid damage to the tissue conditioner and returned immediately to the laboratory (fig. A2.2). The impression is positioned on the working model and plaster keys prepared lingually and buccally around it (figs A2.3 and A2.4). The base and neutral zone impression are then removed from the model and the plaster keys soaked in water or coated with a suitable separator before being replaced on the lower model. Molten wax is poured into the mould space representing the neutral zone. Once the wax has hardened, the plaster keys can be removed. The height of the neutral zone wax replica is checked, and if necessary adjusted such that it occludes with the upper rim at the correct occlusal vertical dimension. Care should be taken not to adjust the buccal and lingual surfaces which represent the limits of the muscular moulding. The teeth should be set

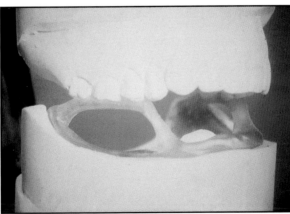

Fig. A2.1 A lower base plate with molar pillars at the correct occlusal vertical dimension on the working model, ready for the neutral zone procedure.

Fig. A2.2 The temporary soft lining material (tissue conditioner) Viscogel, which has been used to record the impression of the neutral zone on a base plate.

complete dentures

Further reading

Basker R M, Davenport J C, Tomlin H R. *Prosthetic treatment of the edentulous patient.* 3rd edn. Basingstoke: Macmillan Press, 1992.

Russell A F. The reciprocal lower complete denture. *J Prosthet Dent* 1959; **9**: 180-190.

up within the confines of the wax (ie the neutral zone) and this should be periodically checked using the plaster keys (fig. A2.5). On occasions, where the buccolingual dimension has been reduced, the posterior teeth may need to be narrowed to keep them within the defined zone (fig. A2.6).

A modification to this technique is the use of an alternative baseplate in the form of an autopolymerising base with wire loops (fig. A2.7) instead of the molar pillars. This technique is identical except that the correct occlusal vertical dimension is not set by the wire loops, as for the molar pillars, but has to be adjusted to the upper occlusal rim or wax trial at the chairside. Other materials may be used to record the neutral zone, for example silicone putty (fig. A2.8), which because of their consistency and relatively rapid setting reaction (although the working time can be extended by reducing the quantity of catalyst in the mix) may, in the view of some clinicians, be easier to manipulate at the chairside but since they are more viscous the intra oral moulding may not be so well defined.

This technique allows the polished surfaces of the lower denture and the artificial teeth to be positioned in an area of minimal muscular activity and reduces displacement of the denture.

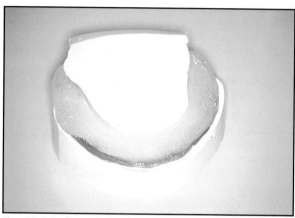

Fig. A2.3 A plaster index has been prepared to define the lingual boundary of the neutral zone.

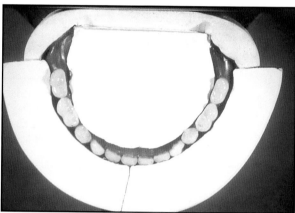

Fig. A2.5 The artificial teeth are positioned within the wax in the neutral zone i.e. within the confines of the plaster indices.

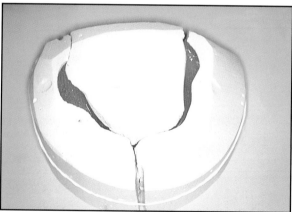

Fig. A2.4 The boundaries of the neutral zone are fully defined by the use of buccal and lingual plaster indices.

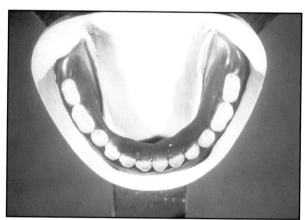

Fig. A2.6 The wax trial illustrates the arch shape as defined by the neutral zone and the narrow width available for the posterior teeth.

complete dentures

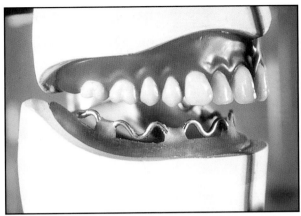

Fig. A2.7 An autopolymerising base plate demonstrating the use of wire loops for the retention of the impression material.

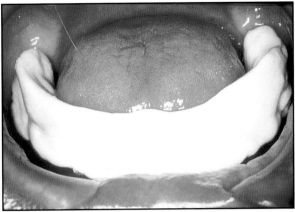

Fig. A2.8 A neutral zone impression utilising silicone putty in the mouth.

Appendix 3: cuspless teeth

The selection of posterior teeth for a complete denture is the responsibility of the dentist but is often delegated to the technician. The choice of artificial posterior teeth is between two types, anatomical and non-anatomical (often referred to as cuspless, monoplane, inverted cusp, zero degree or flat plane). They can be defined as: anatomical — teeth approximating to the forms of human teeth, and non-anatomical — teeth approximating to the forms of human teeth, but having modified occlusal surfaces. Non-anatomical teeth vary with manufacturers, although all have a flat occlusal surface (when viewed from the side) as a common feature but with variations in detail, such as depth of inverted cusp, buccolingual width, the addition of transverse cutters or spillways (fig. A3.1). Mastication occurs by the cusp facets and fissures working in a mortar and pestle like motion.

Some techniques for setting up cuspless teeth

The selection and setting up of cuspless teeth has been described in detail in the dental literature. Various techniques are available and include:

- Free hand arbitary occlusal curvature on an average value articulator.
- 'Sears Technique' also known as ramping the sevens. The teeth are set with a flat occlusal plane and three point contact is achieved by setting the lower second molars at an angle to maintain posterior contact in protrusion. The occlusion is refined when the denture is fitted (fig. A3.2).
- 'Popper's method' which is similar to the above. Occlusal balance is produced by recording the posterior separation of the flat occlusal plane, which is evident in protrusion, at the wax trial stage and resetting the last two posterior teeth on each denture to produce a three point contact.
- Use of metal templates. These may be spherical or sphero-ellipsoid (eg Boyle's plate; fig. A3.3). The technique is similar for both in that the occlusal curvature is generated by setting up the upper teeth to the template and then setting the lower teeth to the uppers. In the spherical plate system all the anterior teeth are set up before positioning and using the template, while with the sphero-ellipsoidal plate, only the two upper central incisors are set up and then, all the other tooth positions are dictated by the plate.
- Plaster and pumice rims. The centric jaw registration is completed as for routine denture construction, but using wax occlusal rims with acrylic resin bases. The shaped rims are then copied in a 50:50 mix of plaster of Paris and pumice on to the acrylic resin bases. The occlusal vertical dimension of these rims is made 3 mm greater than that recorded for the individual. At the next clinical visit the individual then generates his or her own occlusal plane by gently grinding the rims together in the mouth, making eccentric movements to all border positions. The upper teeth are then set to the lower rim, followed by setting the lower teeth to the uppers (fig. A3.4). This can be a tedious and somewhat messy clinical procedure.

In all cases the anterior teeth must be set with no vertical overbite unless there is a sufficiently large horizontal overjet to permit protrusion without anterior tooth contact.

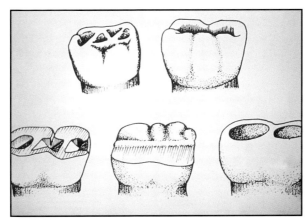

Fig A3.1 Examples of the occlusal surfaces of a variety of cuspless teeth.

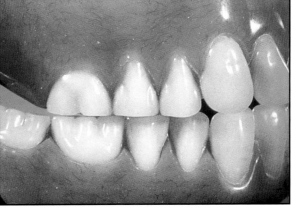

Fig A3.2 Cuspless posterior teeth set to a flat occlusal plane with the last molar ramped in order to obtain anterior and posterior contact in protrusion.

complete dentures

Further reading

Bates J F, Huggett R, Stafford G D. *Removable denture construction.* pp 47-50. Oxford: Wright, Butterworth Heinmann, 1991.

Harrison A, Huggett R. Cusped vs cuspless teeth — the choice is yours. *Dent Pract* 1977; **15**: 9-12.

Huggett R. Techniques used for setting up cuspless posterior teeth for complete dentures. *Dent Technician* 1970; **23**: 4-8.

Advantages of cuspless teeth

Proponents of cuspless teeth suggest their use can eliminate displacing cuspal contacts and lateral stresses on the underlying denture bearing tissues and recommend their use in specific situations, for example:

- severe mucosal atrophy and extensive ridge resorbtion. A cuspless scheme eliminates the possibility of deflective occlusal contacts and may increase the lateral stability of the dentures and reduce trauma.
- difficulty in recording centric jaw registration in the elderly or due to an underlying neuromuscular problem such as Parkinson's disease;
- when there are abnormal jaw relationships their use may permit closure and contact over a broader area;
- replacement dentures with pre-existing cuspless teeth;
- replacement dentures by a duplication technique;
- a combination of the above.

Also, as there is no interdigitating occlusal form setting up should require less laboratory time.

Disadvantages of cuspless teeth

Opponents however suggest disadvantages in their use, for example:

- in the absence of cusps it may not be possible to chew food efficiently. It has been suggested in the literature that cusped teeth provide the best performance but this is also related to the retention and stability of the denture bases. Certainly some form of occlusal markings or cuspal indentations is more efficient than purely flat tooth forms;
- occlusal loads may be sustained for prolonged periods. However, this is unlikely since cuspless teeth tend to be narrower occlusally;
- some clinicians feel that cusps produce a 'centric stop' which gives the patient a more comfortable and positive end point to the jaw closing cycle;
- poor aesthetics. The transition between the naturally shaped canine and the flat occlusal surface of the first premolar may be obvious.

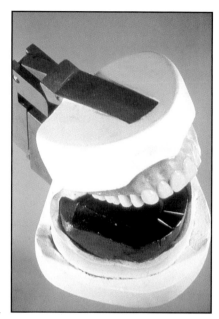

Fig. A3.3 An examples of cuspless teeth set to a Boyle's Plate using a simple hinge articulator.

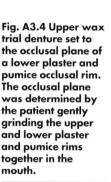

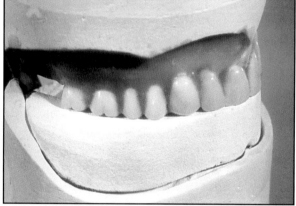

Fig. A3.4 Upper wax trial denture set to the occlusal plane of a lower plaster and pumice occlusal rim. The occlusal plane was determined by the patient gently grinding the upper and lower plaster and pumice rims together in the mouth.

Appendix 4: copy dentures

Many denture wearers, especially the elderly, who require replacement dentures, may have been wearing the same set of dentures over a long period of time. Complete dentures are controlled by the oral musculature mainly through the polished surfaces, a skill learnt by habituation. The capacity for adaptation is reduced in the elderly and they can have difficulty in controlling new dentures which are different in shape from the existing set. This particularly applies to the contours of the polished surfaces. It is possible to reproduce the polished surfaces of the existing dentures together with the introduction of selective modifications by using a copy denture technique. This technique has also been referred to as the duplicate denture technique, however strictly speaking this terminology should only refer to a true duplicate of the original denture where no modifications are made. The aim of the copy technique is to utilise the desirable features of the existing dentures but to improve the other aspects.

Indications for the use of a copy denture technique

- If the existing complete dentures have satisfactory occlusal, polished and impression surfaces and the individual requests a spare set.
- If modifications are required to existing dentures to improve the impression surface and/or the occlusal surface but the polished surfaces are satisfactory.
- For individuals who possess multiple sets of complete dentures. If the set which is usually worn is the 'best of a bad bunch' they can be used as a template for modifications.
- For replacement dentures where there has been general deterioration of the denture base material.

Contraindications for the use of a copy denture technique

- If the polished surfaces of the present denture are incorrect, ie not in the neutral zone.
- Obviously, if the previous dentures have been lost or are not available the technique cannot be used.

Technique

The choice of technique will depend on the availability of laboratory support, the accuracy of the copy required and personal choice. The aim is to form replicas of the existing dentures which can be used as templates during the construction of complete dentures. A simple technique, suitable for use in general dental practice is described.

Four upper plastic disposable stock impression trays are selected and silicone adhesive applied to the inner surfaces of two and the

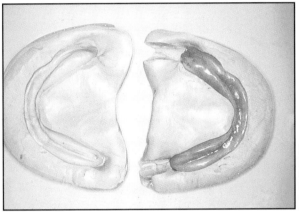

Fig. A4.1 Polysiloxane putty copy impression of a complete lower denture.

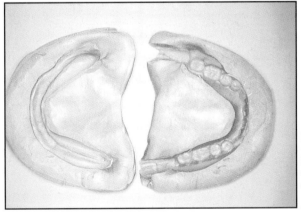

Fig. A4.2 Denture removed from copy impression and a check is made of the accuracy of surface reproduction.

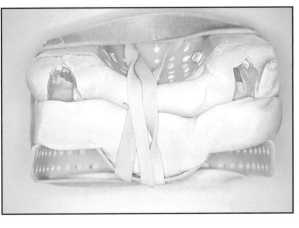

Fig. A4.3 The relocated halves of the copy impression secured with elastic bands. The sprue holes are clearly visible.

complete dentures

outer surface of the other two trays. The dentures to be copied are cleaned, disinfected and dried. An appropriate amount of polysiloxane impression putty is mixed and placed in one of the trays with the adhesive on the inner surface. The denture is inserted occlusal surface down, and putty manipulated to within 2 mm of the periphery. When set, petroleum jelly is applied to the exposed putty, for ease of separation of the two halves of the mould.

A second mix of putty is then manipulated into the fit surface of the denture and an impression tray with adhesive on its outer surface, positioned on top of the putty. The residue is wrapped around the tray. When set, the mould is separated, a scalpel is used to prepare access (sprue) holes in the posterior aspect of the mould and the denture removed (figs A4.1 and A4.2). The mould halves can then be secured together with elastic bands whilst autopolymerising acrylic resin is mixed in a laboratory fume cupboard or a suitably vented cabinet (fig. A4.3). The fluid acrylic resin is poured into the access holes and the sides of the mould tapped to disperse air bubbles. To ensure adequate curing each mould is placed in a hydroflask containing warm water for 15 mins.

When set the copy templates are removed from the mould and checked for errors, for example, positive blebs, air blows or voids (fig. A4.4). The copies are trimmed with an acrylic bur to remove excess material in the region of the sprues, polished and are then ready for use (fig. A4.5).

Modifications of this technique include the use of alginate impression material instead of polysiloxane impression putty and the use of an alternative container such as a modified soap box (figs A4.6 and A4.7). Molten wax can be used for the reproduction of the teeth instead of autopolymerising acrylic resin, which makes it easier for the technician to remove and replace with artificial teeth when setting up (fig. A4.8). It is unwise to use wax for the entire template since in an unreinforced state it is likely to distort.

Clinical stages

The templates are effectively used as special trays. They are modified with greenstick impression tracing compound to correct underextension and to ensure properly moulded peripheries. Working impressions are taken in a light bodied silicone or zinc oxide eugenol impression material. The jaw relationship can then be recorded together with any modifications to the occlusal vertical dimension, with wax or bite registration paste and a tooth shade taken.

The impressions are cast in the laboratory and the templates mounted on an articulator. The technician should remove each tooth individually from the template and replace with an artificial tooth, to ensure as accurate a reproduction as possible (fig. A4.9). The copies are returned to the clinic for the trial stage and finished as for routine complete dentures (fig. A4.10). It should be noted that the autopoly-

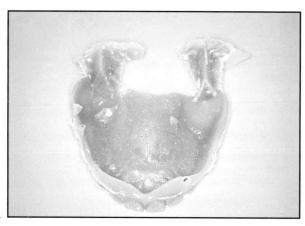

Fig. A4.4 A poor autopolymerising acrylic resin copy of the original denture demonstrating porosity. Some areas of the copy show obvious excess whereas in others there are deficiencies.

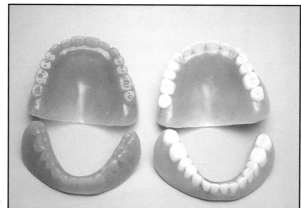

Fig. A4.5 Trimmed and polished copy templates alongside the original dentures.

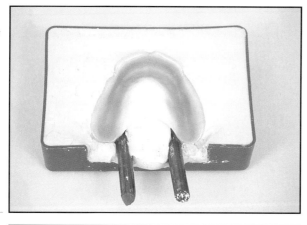

Fig. A4.6 A plastic soap dish, modified to provide a window for sprue access, is used as a container for an alginate impression. Greenstick compound is added to form the sprue holes before the denture is inserted into the impression material. When set a second mix of alginate is used for the other half of the mould.

Fig. A4.7 The denture has been removed from the mould and the halves relocated and fluid autopolymerising resin is carefully poured through a sprue hole.

complete dentures

merised bases used in the clinical stages are discarded during the final laboratory processing and the new dentures are constructed using conventional heat cured acrylic resin.

Further reading

Cooper J S, Watkinson A C. Duplication of full dentures. *Br Dent J* 1976; **141**: 344-348.

Davenport J C, Heath J R. A copy denture technique. *Br Dent J* 1983; **155**: 162-163.

Duthie N, Lyon F F, Sturrock K C, Yemm R. A copying technique for replacement of complete dentures. *Br Dent J* 1978; **144**: 248-252.

Heath J R, Johnson A. The versatility of the copy denture technique. *Br Dent J* 1981; **150**: 189-193.

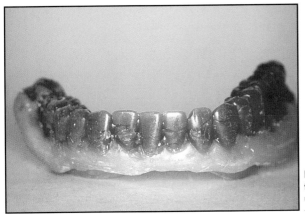

Fig. A4.8 An example of a copy template where wax has been used to represent the teeth.

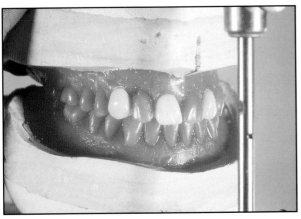

Fig. A4.9 Alternate teeth are removed from the template and replaced with artificial teeth. Replacement in this way ensures accurate reproduction of tooth position and relationship to adjacent teeth.

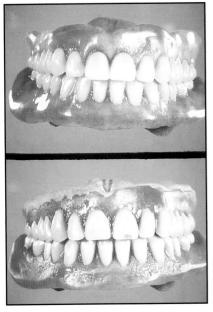

Fig. A4.10 The completed copy dentures alongside the originals to demonstrate the value of the technique.

Appendix 5: denture fixatives

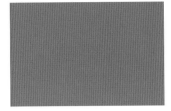

Following the loss of the natural dentition the edentulous individual can be faced with a number of problems associated with wearing complete dentures. One of the most common is that of looseness of either or both of the dentures. Denture fixatives are a group of materials which can be used to aid the retention of dentures. Some clinicians disapprove of their use believing that they may be used with old dentures instead of having a properly constructed accurately fitting new denture. It is suggested that their inappropriate use can lead to problems such as impairment of the development of the neuromuscular system which is important for the retention and stability of dentures.

Other workers accept that denture fixatives have a valuable function in certain situations and contribute significantly to denture retention and stability. Which ever of these views is taken is perhaps hardly relevant since, although it has been suggested that these products should only be available on prescription, the denture wearer is able to purchase them over the counter.

For those individuals for whom retention is a major difficulty the regular use of fixatives becomes the norm. They may be self prescribed to improve the retention of clinically inadequate dentures, for enhanced comfort or simply for the added sense of security that comes with their use.

Types and constituents of denture fixatives

Fixatives are marketed in various formulations and can broadly be divided into two groups. The first group are 'temporary' in the sense that they are eventually lost from the denture and inevitably swallowed during normal wear along with food and drink. This group includes those marketed in the form of creams, powders and liquids. The second group is supplied in preformed sheets which can be trimmed to the desired shape and are 'permanent' in that they contain insoluble products (eg paper) although their lifespan is usually determined by the level of expertise used in their application to the denture base. It has been suggested that the introduction of a non displaceable material between the denture and supporting tissues can result in soft and hard tissue damage.

The constituents of the first group can be divided into the following categories:

- adhesive components eg gelatin, pectin, sodium carboxymethylcellulose
- antimicrobial agents eg sodium borate, sodium tetraborate, hexachloraphene
- fillers eg magnesium oxide, sodium phosphate, calcium silicate
- flavouring agents eg peppermint oil
- wetting agents eg mineral oil
- preservatives.

Most manufacturers of denture fixatives are reluctant to reveal the exact formulation of their products. However, for information, the approximate formulations of some popular denture fixatives according to their manufacturer are given in Table A5.1.

Mechanism of action

The mode of action of denture fixatives has been described in the dental literature and it has been reported that the effectiveness of the fixatives is dependent on both chemical and physical forces. The adsorption of water and saliva by fixatives results in the formation of a layer which obliterates the space between the denture

Table A5.1 Some popular denture fixatives and their constituents according to the manufacturer

Type	Manufacturer	Constituents
Fixodent (cream)	Procter and Gamble Ltd Egham, Surrey, UK	60% calcium sodium poly(vinyl methyl ether maleate) 40% sodium carboxymethylcellulose
Fixodent (powder)	" " "	" " "
Boots denture fixative (cream)	The Boots Company PLC Notts, UK	Cellulose gum Petrolatum Mineral oil Peppermint oil Butyl paraben Methyl paraben Menthol Eucalyptus oil BHT CI 45430
Boots denture fixative (powder)	" "	Cellulose gum
Poli-grip Ultra (cream) Super Poli-grip (cream) New Wernets Ultra (powder) Super Wernets (powder) Dentuhold (liquid)	Stafford Miller Ltd Welwyn Garden City Herts, UK	Thickening Suspending agents Adhesive agents Colouring agents Preservatives Flavouring agents

Further reading

Chew C L, Boone M E, Swartz M L, Phillips R W. Denture adhesives: their effect on denture retention and stability. *J Dent* 1985; **13**: 152-159.

Ellis B, Al-Nakash S, Lamb D J. The composition and rheology of denture adhesives. *J Dent* 1980; **8**: 109-118.

Gates W D, Goldschmidt M, Kramer D. Microbial contamination in four commercially available denture adhesives. *J Prosthet Dent* 1994; **71**: 154-158.

Ghani F, Picton D C A, Likeman P R. Some factors affecting retention forces with the use of denture fixatives *in vivo*. *Br Dent J* 1991; **171**: 15-21.

Hogan W J. Allergic reactions to adhesive denture powders. *NY State Dent J* 1954; **20**: 65-66.

Jagger D C, Harrison A. Denture fixatives — an update for general dental practice. *Br Dent J* 1996; **180**: 311-313.

Ow R, Bearn E. A method of studying the effect of adhesives on denture retention. *J Prosthet Dent* 1983; **50**: 332-337.

Polyzois G. An update on denture fixatives. *Dent Update* 1983; October, 579-583.

Shay K. Denture adhesives — choosing the right powders and pastes. *JADA* 1991; **122**: 70-76.

Stafford G D. Denture relining material for home use. Report of a case. *Br Dent J* 1973; **134**: 391-392.

Tarbet W J, Boone M, Schmidt N F. Effect of denture adhesive on complete denture dislodgement during mastication. *J Prosthet Dent* 1980; **44**: 374-378.

Tarbet W J, Silverman G, Schmidt N F. Maximum incisal biting force in denture wearers as influenced by adequacy of denture bearing tissues and the use of an adhesive. *J Dent Res* 1981; **60**: 115-119.

Woefel J, Winter C, Curry R. Additives sold over the counter dangerously prolong wearing period of ill fitting dentures. *JADA* 1965; **71**: 603-613.

base and the mucoperiosteum. The thickness of the layer is of importance as too thick a layer will lead to a decrease in retention. The effectiveness of fixatives has been explained in terms of anionic/cationic interaction whereby water adsorption of the adhesive agents leads to attraction between anions and cations in the mucous membrane accounting for stickiness. An increase in viscosity of the fixative layer leads to an increase in retention.

Denture retention, stability and masticatory efficiency

There is a wealth of information to support the statement that denture fixatives contribute to denture retention and stability. An increase in biting force associated with the use of denture fixatives has also been demonstrated. An increase in the occlusal vertical dimension may be noted with the use of denture fixatives, however this should be minimal if the fixative is used correctly and in which case the occlusion should not be affected. Substantial changes may be seen however with the inappropriate use of multi layers of preformed sheets.

Biological aspects of denture fixatives

With time fixative is lost from the denture and inevitably swallowed along with food and drink. This loss over time is recognised but despite the potential long term ingestion of fixatives and their leachable components there have been few reports on this aspect. One constituent in particular (karaya gum) has been linked to possible side effects of nausea and epigastric pain along with allergic reactions such as angionuerotic oedema and hives.

Microbiological aspects of denture fixatives

Denture fixatives often include antimicrobial agents such as hexachloraphene, sodium tetraborate and ethanol. Their long term use may influence the oral microflora by selectively supporting the growth of some microorganisms and inhibiting others. However, some reports claim that fixatives do not possess inhibitory properties towards micro-organisms. More recently it has been suggested that the use of denture fixatives should be limited in the immunocompromised patient, as some can contain bacterial and fungal contaminants that may cause infection in these individuals.

Recommendations for the use of denture fixatives

Denture fixatives can be recommended for use and to aid retention in the following situations:
- Those with poor anatomical denture support who have continued difficulty with retention although the dentures are clinically and technically satisfactory;
- individuals with poor neuromuscular control; (eg stroke and Parkinsonism)
- xerostomia sufferers (eg Sjogren's syndrome, post radiotherapy, drug induced)
- obturators.

Denture fixatives may have some value for a limited period:
- Following the insertion of new dentures when difficulties are being experienced in developing neuromuscular control.

Denture fixatives should not be used in the following situations:
- Following the insertion of immediate dentures to ease post operative discomfort - use of a temporary lining material such as Viscogel (Dentsply Ltd, Weybridge, UK) or Coecomfort (Coe Laboratories Inc, Chicago, USA) is the treatment of choice;
- for the retention of clinically unacceptable dentures.

Application of denture fixatives

Most manufacturers provide step by step instructions, often with diagrams, for the application of fixative to the fitting surface of the dentures. It is generally recommended that the dentures are rinsed with water prior to application. For the fixatives in paste/cream form it is recommended that it is applied in short strips to the fitting surface of a dry denture. Those in powder form should be applied in a thin layer to the fitting surface of a moist denture. Denture fixatives in liquid form should be applied as a fine line to the fitting surface of a dry denture.

For maximum effect the denture is held firmly in place for a few seconds and eating or drinking should be avoided for a period immediately after the application of fixative.

Removal of denture fixatives

It is generally recommended that dentures should be removed before retiring. Denture fixative should therefore be removed from the denture by brushing with a soft brush and warm water before the dentures are cleaned by an approved method and stored in water over night.

Denture fixatives are readily available for purchase over the counter. If used inappropriately they can cause or mask underlying problems but they can be of benefit if their use is limited to specific recommended situations. The dentist should have a working knowledge of the types of denture fixatives available, be able to give informed advice on their advantages and disadvantages and make recommendations for their use.

Appendix 6: soft lining materials

Soft lining materials can be defined as soft, resilient, elastic materials which form a cushioned layer between the hard denture base and the oral mucosa. They are broadly divided into two groups of materials. The first group is the temporary soft materials and includes tissue conditioners typically based on poly (ethyl methacrylate), an aromatic ester and ethyl alcohol. As the name suggests these materials are indicated for use in temporary or transitional situations. The second group includes the permanent soft lining materials based on silicone rubber or acrylic resin. The use and misuse of products available over the counter for home use for self relining the fitting surface of removable prostheses and the subsequent damage to the denture bearing areas is well documented and these materials will not be discussed further.

Temporary soft materials

Constituents

Temporary soft materials can be divided into tissue conditioners or temporary soft lining materials, although manufacturers market the same products for both purposes. To fulfill these requirements the materials need to display viscoelastic properties in that they should flow under steady pressure but be resilient under dynamic forces such as chewing. In general they are supplied in powder and liquid form. The powder is usually poly (ethyl methacrylate) and the liquid contains an aromatic ester and ethyl alcohol. However the composition varies between products. A list of some popular temporary soft materials together with their constituents and manufacturers is given in Table A6.1.

Indications for use of temporary soft materials

Temporary soft materials are useful materials if used with a degree of expertise in the appropriate situations. The indications for their use have been discussed in detail in the dental literature. They can be used as tissue conditioners or as temporary linings. The powder and liquid are mixed together to form a slurry. The setting reaction is a physical, rather than a chemical, process with the liquid penetrating the powder particles to form a gel. The material only remains soft for a limited period of time as subsequent leaching and evaporation of some of the components leads to hardening of the gel, and as a consequence, frequent replacement of the lining may be necessary. Although these materials are only designed to be used for a few days, they can last a few weeks if correctly applied and not damaged by incorrect cleaning techniques. However, they should not be left for indeterminate periods since they will become hard and rough and, in some cases, stained and foul. The plasticiser component of temporary soft materials can have an adverse softening effect on acrylic resin, particularly so for auto polymerised acrylic resin, although it is unlikely that the effect on heat polymerised acrylic resin is significant.

Tissue conditioners

Trauma. A tissue conditioner is a soft material which is applied temporarily to the fitting surface of a denture in order to allow a more even distribution of load to the denture bearing area and promote resolution and healing of the tissues. Any occlusal errors, pressure areas and incorrect denture base extensions should be rectified before tissue conditioners are used. The use of a tissue conditioner coupled with improved oral hygiene may be indicated, prior to beginning the construction of a new prosthesis, when there is inflammation of the denture bearing area. When mixed the material should be applied to the entire fitting surface of clean, dry dentures. The dentures should be inserted and gently brought into occlusion for approximately 2 minutes. After border moulding the dentures are removed and excess material removed with a scalpel or hot wax knife, taking care to protect the moulded peripheries. After approximately 5 days all traces of tissue condi-

Table A6.1. Some popular temporary soft materials together with their constituents (where available) according to the manufacturer

Temporary soft material	Constituents	Manufacturer
Viscogel	Poly (ethyl methacrylate) Ethyl alcohol Dibutyl Phthalate	De Trey Division Dentsply Ltd Weybridge UK
Coe Comfort	Poly(methyl methacrylate) Ethyl alcohol Dibutyl Phthalate Zinc oxide undecylenate	Coe Laboratories inc Illinois USA
Dinabase	Not available from manufacturer; single component, high viscosity monomer free	Quattroti Dentech Italy

complete dentures

Further reading

Bauman R. Chairside modification of dentures for tissue conditioning materials. *J Prosthet Dent* 1978; **40:** 225-226.

Braden M. Tissue conditioners 1: Composition and structure. *J Dent Res* 1970; **000:** 145-148.

Davenport J C, Wilson H J, Spence D. The compatibility of soft lining materials and denture cleansers. *Br Dent J* 1986; **161:** 13-17.

Davis M, Carmichael R. The plasticising effect of temporary soft lining materials on polymerised acrylic resin. *J Prosthet Dent* 1988; **60:** 463-466.

Douglas W H, Walker D M. Nystatin in denture liners — an alternative treatment of denture stomatitis. *Br Dent J* 1973; **135:** 55-59.

Goll G, Smith D E, Plein J B. The effect of denture cleansers on temporary soft liners. *J Prosthet Dent* 1983; **50:** 466-472.

Graham B S, Jones D W, Burke J, Thompson J P. In vivo fungal presence and growth on two denture liners. *J Prosthet Dent* 1991; **65:** 528-532.

Harrison A. Temporary soft lining materials. A review of their uses. *Br Dent J* 1981; **151:** 419-422.

Harrison A, Basker R M, Smith I S. The compatibility of temporary soft lining materials with immersion denture cleansers. *Int J Prosthodont* 1989; **2:** 254-258.

Jagger D C, Harrison A. Denture cleansing — the best approach. *Br Dent J* 1995; **178:** 413-417.

tioner should be removed from the denture and replaced with fresh material, if necessary. This treatment and reinforcement of oral hygiene should be continued until the inflammation subsides.

Immediate dentures. Temporary soft material may be used as a temporary reline for an immediate denture as an interim measure until sufficient post extraction resorption has occurred and a permanent reline can be performed. There is no definitive time as to when is best to permanently reline an immediate denture however, as a guideline, it has been suggested that 40% of post-extraction changes in the maxilla have occurred by the end of the first month (expressed as a percentage of average changes over 30 months); 65% by the end of the third month and 80% by the end of six months. There is therefore very little point in permanently relining a denture until the early, rapid post-extraction changes have occurred. The material is applied to the fitting surface of the dentures as previously described. It can be left *in situ* for longer than when used as a tissue conditioner but should be replaced when it shows evidence of becoming roughened and hard.

Denture-induced stomatitis. Denture-induced stomatitis is a common condition reported in patients wearing complete dentures. It often presents as a patchy diffuse inflammation of the mucosa covered by the denture (almost always the upper) and is associated with poor oral hygiene and denture plaque, *Candida albicans* and trauma from dentures. The tissue conditioner is applied to the fitting surface of the denture as previously described but treatment should be coupled with denture hygiene and appropriate anti fungal therapy or the use of a tissue conditioner with an incorporated anti fungal component may be considered. Ideally, prior to the use of an antifungal agent the presence of a yeast should be established by culture, however this is not always possible in general practice.

Functional impressions

A functional impression may be described as an impression achieved under functional stresses.

Fig. A6.1 A temporary soft material (tissue conditioner) applied to a gutta percha bung on an obturator to accurately record a functional impression of an intra oral defect.

The properties required of a functional impression material are different than for tissue conditioning. It should have the ability to flow but not recover ie retain its shape when the impression is completed and during model casting. Although not ideal in theory, materials used as tissue conditioners have been used successfully as functional impression materials.

To record a functional impression the material is added to the fitting surface of the denture as for tissue conditioning purposes, again ensuring prior correction of any occlusal irregularities or border over extension. The denture should be worn for a period of normal function. Although there is some controversy as to what is regarded as an acceptable length of time for this purpose, a period of 4–6 hours has been shown to be acceptable. The denture is then removed and the impression cast immediately to avoid distortion. Functional impressions are particularly valuable in the treatment of patients with congenital oral defects, such as cleft palate, where they can be used to achieve a functional record of the intraoral defect for the final shaping of the obturator (fig. A6.1). Alternatively they can be used in post surgical situations to enhance the fit and seal of a prosthesis and have the advantage of being easy to modify as the tissues heal and the intraoral defect contracts. These materials can be used long-term, with appropriate replacement, until a permanent prosthesis is made.

Maintenance of temporary soft materials

For maximum resolution of inflammation in the tissues, the material needs to be soft and plastic, ie it should flow under load, and therefore is susceptible to damage. Inadequate or incorrect cleaning of these materials can result in a more rapid deterioration, with discolouration, malodour, hardening and subsequent irritation to the oral mucosa. The correct combination of cleanser and temporary soft material is essential. Brushing has a damaging abrasive effect and should be avoided as should the use of effervescent peroxides which can cause bubbling and deterioration of the surface. It is recommended that dentures with temporary soft materials are rinsed in water after each meal and cleaned daily by soaking for 20 minutes in an alkaline hypochlorite solution. However the use of hypochlorite can cause moderate surface damage to Coe Comfort, and possibly other materials, after 14–21 days use.

Permanent soft lining materials

Constituents

The second category of materials is comprised of those intended as permanent soft linings and should not be confused with the temporary group discussed earlier. Permanent soft lining materials are long-term resilient materials attached to the fitting surface of a denture forming a cushion between the denture and the underlying tissues. Unfortunately they are frequently used as a 'soft option' blanket treatment for problems such as chronic pain under a lower denture, when it would be more appropriate to diagnose and eliminate any possible precipitating factors such as occlusal or fitting surface errors or incorrect denture base extension. Using this method to solve problems can bring difficulties in itself in that the denture wearer may become accustomed to a soft lining and find difficulty in re-adapting to a conventional hard denture base.

A list of some popular permanent soft lining materials together with their constituents and manufacturers is given in Table A6.2. They can broadly be divided into two groups depending on their constituents, namely plasticised acrylic resin and silicone rubber. These in turn can be subdivided into heat and auto polymerised. The acrylic resin group are presented in powder/liquid form. The powders usually consist of beads of polyethyl- or polybutyl methacrylate and the liquid is either methyl or butyl methacrylate monomer and a plasticiser, for example dibutyl phthalate. The silicone rubbers are generally polymers of dimethylsiloxane together with cross linking agents, for example acryloxyalkysilane and a catalyst such as benzoyl peroxide. Unlike the acrylic resin variety there is no direct bond between the denture base and the soft lining and an adhesive is required, often in the form of a silicone polymer.

Indications for use of permanent soft lining materials

Atrophic/knife edge alveolar ridge. Gradual resorption of the mandibular alveolus with age leads to a reduction in the denture bearing area and often a knife edged atrophic ridge. The mucosa can be thin and easily traumatised by the overlying hard denture base, particularly so in those individuals with clenching or grinding habits. A soft lining added to a fully extended base may be helpful in absorbing and disseminating some of the occlusal loads in these circumstances.

Superficial mental nerve. In some individuals the gradual resorption of bone can lead to the mental nerve lying on the surface of the alveolar ridge. The nerve can be trapped between the denture and the tissues causing pain. Inclusion of appropriate relief on the master cast together with a soft lining may help to relieve the pressure on the nerve.

Extensive bony prominences. A soft lining may be used overlying prominent maxillary and mandibular tori or mylohyoid ridges where the overlying mucosa is often thin and can be easily traumatised. Although this is advocated in textbooks it is not a widely adopted treatment option, as it increases the bulk of the denture, especially in the mid palatal region over maxillary tori. An alternative approach is simply to relieve the denture over the prominent areas.

Congenital/acquired oral maxillofacial defects. Soft linings can be used for congenital or acquired oral defects such as with obturators for cleft palate patients, allowing the utilisation of undercuts for retention of the prosthesis which may not otherwise be engaged with hard acrylic resin.

Xerostomia. Soft linings have been recommended for those with a reduced saliva flow, possibly as a result of degenerative changes in the salivary glands, radiotherapy or drug therapy. Adequate saliva is of benefit to the denture wearer both for lubrication and to aid retention and its absence can result in loose dentures which cause discomfort and soreness. However, although this approach may be recommended in textbooks, the use of the silicone rubber type of soft linings with their lack of wettability, can result in an increase in trauma to the mucosa due to friction if the denture is loose and is dragged across the tissues. Also, candidal colonisation of permanent resilient linings is not an uncommon feature in the edentulous patient with a dry mouth.

Limitations of use of permanent soft lining materials

Fracture of the denture base. In order for the soft lining material to act as a cushion it must be of adequate thickness, normally at least 2–3 mm.

Jepson N J A, McCabe J F, Storer R. Age changes in the viscoelasticity of permanent soft lining materials. *J Dent* 1993; **21:** 171-178.

Loney R W, Moulding M B. The effect of finishing and polishing on surface roughness of a processed resilient denture liner. *Int J Prosthodont* 1993; **6:** 390-396.

Nikawa H, Iwanga H, Hamada T, Yuhta S. Effects of denture cleansers on direct soft denture lining materials. *J Prosthet Dent* 1994; **72:** 657-662.

Qudah S, Harrison A, Huggett R. Soft lining materials in prosthetic dentistry. A review. *Int J Prosthodont* 1990; **3:** 477-483.

Stafford G D. Denture relining material for home use. Report of a case. *Br Dent J* 1973; **134:** 391-392.

Watson R M. The use of tissue conditioners for obturator impressions. *Br Dent J* 1968; **124:** 226.

Wilson H J, Tomlin H R, Osborne J. Tissue conditioners and functional impression materials. *Br Dent J* 1966; **121:** 9-16.

Wright P S. Composition and properties of soft lining materials for acrylic dentures. *J Dent* 1981; **9:** 210-223.

Wright P S. Observations on long term use of a soft lining material for mandibular dentures. *J Prosthet Dent* 1994; **72:** 385-392.

Table A6.2. Some popular permanent soft lining materials together with their constituents (where available) according to the manufacturer

Material	Constituents	Manufacturer
Molloplast B	Poly(dimethylsiloxane) Acryloxyalkylsilane Heat + Benzoyl peroxide y-methacryloxy propyl- trimethoxysilane	Austenal Harrow, Middlesex UK
Flexibase	Poly(dimethylsiloxane) Triethoxysilanol Dibutyl tin dilurate Silicone polymer in solvent	Flexico Developments London, UK
Eversoft	Poly(ethyl methacrylate) Dibutyl phthalate Ethyl acetate Ethyl alcohol Methyl ethyl ketone	Austenal Harrow, Middlesex UK
Coe - Soft	Poly(ethyl methacrylate) Di-n-butyl phthalate Benzyl salicylate Ethyl alcohol	Coe Laboratories Illinois, USA
Tota	lLong chain acrylic polymers and monomers	Stratford Cookson Company New York, USA
Tokuyamata	Silicone based Adhesive primer	Tokuyama Corp Tokyo, Japan
Vertex	Cadmium free Poly(ethyl methacrylate) Acetyl tributylcitrate Methyl methacrylate	Dentimex Netherlands

complete dentures

This often necessitates a compensatory reduction in thickness of the denture base material to minimise an increase in the occlusal vertical dimension. For denture bases of less than optimum thickness, possibly due to limited inter ridge space, fracture is a frequent problem. The use of high impact acrylic resin or a reinforcement, possibly in the form of a cast cobalt chromium lingual plate, will reduce this problem, although these can be problematic (fig. A6.2). It is not uncommon for those whose dentures have been supplied with a soft lining to complain of persistent pain. This is often because the soft linings are of inadequate thickness and/or the bases are underextended with a resultant greater pressure on the limited area of support.

Adhesion. A common finding is failure of adhesion between the silicone soft linings and the denture base resulting in 'peeling off' of the soft lining (fig. A6.3). A poor laboratory procedure may be the cause although rough handling at an unsupported junction is a common culprit. A boxed in lining, where the soft lining does not include the border area of the denture, probably has the longest lifespan. However boxed in linings should be restricted to use where pain is limited to the crest of the ridge and not where pain is related to the border.

Adjustments. Permanent soft linings are difficult to trim, finish and polish, often producing a roughened surface which can traumatise the oral mucosa or hasten the accumulation of plaque. Methods for finishing and polishing have been reported in the dental literature and recommendations made. For example, for Molloplast B it is recommended that initial reduction is carried out with a Brasseler 747-140 green stone, a Brasseler 747-340 pink stone(Brasseler, Montreal, Canada), or a Prolastic trimming wheel (Prolastic, Young Dental, St Louis, USA). For subsequent polishing a dry rag wheel with tin oxide can be used. If adjustments are necessary minimal force and slow speeds are advocated to avoid overheating and subsequent tearing of the soft lining.

Longevity. The longevity of permanent soft linings is a major consideration. Their usefulness is undisputed but their deterioration with time and likely need for replacement means they are often only regarded as semipermanent and their use discouraged. However, favourable longevity results over a 9 year period have been reported for a popular permanent soft lining (Molloplast B).

Maintenance of permanent soft lining materials
Permanent soft linings can be colonised by Candida albicans and if not cleaned thoroughly can become stained and harbour odours (fig. A6.4). However care must be taken in choosing an appropriate method of denture hygiene as inappropriate cleaning regimens can cause hardening and / or bleaching of the soft lining.

For both silicone and acrylic resin materials similar cleaning procedures are recommended as for temporary linings. The denture should be rinsed after every meal and debris removed by brushing with a soft brush, soap and cold water. The denture should be soaked in an alkaline hypochlorite solution for 20 minutes in the evening, be rinsed thoroughly with cold water and soaked in cold water overnight. The other main types of denture cleanser available for purchase over the counter can have mild effects on soft linings such as loss of colour from immersion in acidic cleansers.

This appendix has provided an update on temporary and permanent soft materials and guidance on their indications and limitations of use. With limitation of their use to recommended situations and careful maintenence the 'soft option' can fulfill a valuable role in prosthetic dentistry.

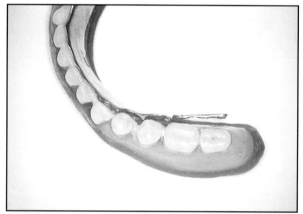

Fig. A6.2 A lower complete denture with a permanent soft lining which has been reinforced with a cobalt chromium plate. There has been loss of attachment between the plate and acrylic resin denture base material.

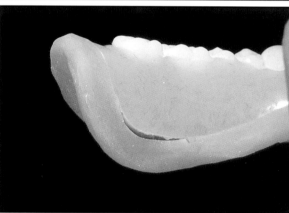

Fig. A6.3 The unprotected border of the permanent soft lining material has pulled away from the denture base.

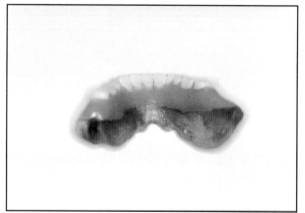

Fig. A6.4 A stained and discoloured permanent soft lining material on a lower complete denture.

Appendix 7: Medical Devices Directive

The Medical Devices Directive (Directive 93/42/EEC) was formally adopted by the European Council of Ministers in June 1993. The Directive regulating the safety and marketing of all medical devices, whether used in the public or private sector, requires member states of the European Community to put into effect provisions to implement the Directive as from January 1995 but allowed a transitional period when manufacturers could choose to follow existing national regulations up to June 1998.

The Directive has two major aims: to ensure that medical devices placed on the market do not compromise the health and safety of patients, users and, where applicable, any other person; to ensure that devices achieve their intended performance for quality and safety.

From 13 June 1998 all dental laboratories manufacturing and placing dental appliances (regarded as custom made devices) on the market have had to comply with requirements of the Medical Devices Directive and be registered with the Medical Devices Directive Agency which is an executive agency of the Department of Health. It is important therefore that dentists who manufacture appliances for their patients, or who own a dental laboratory, understand their responsibilities. To comply with the Medical Devices Directive a quality control system for manufacture has to be established in a laboratory and registered with the Medical Devices Agency. Within the quality control system the following criteria have to be met:
- a method of traceability of materials used by the laboratory
- a method of accountability and traceability of who made an appliance for a patient and when
- a method and record of certification for finished products
- documentation of quality standards used for the manufacture of devices.

All the documentation associated with the manufacturing process, for example the laboratory work cards, materials records etc., have to be stored (for 5 years) but be readily available for inspection by the Agency and the manual of quality assurance and compliance plus any relevant generated audits must also be available.

Definitions
In order to understand the Directive certain definitions are required.

- *Manufacturer* — the natural or legal person with responsibility for the design, manufacture, packaging and labelling of a device before it is placed on the market under his own name, regardless of whether these operations are carried out by that person himself or on his behalf by a third party. The Directive states that in the manufacturing cycle of a dental appliance it is the dentist who designs the product and the dental laboratory manufactures it to a predefined specification.
- *Custom made device* — any device specifically made in accordance with a duly qualified dental (or medical) practitioner's written prescription which gives, under his responsibility, specific design characteristics and is intended for the sole use of a particular patient. The above mentioned prescription may also be made out by any other person authorised by virtue of his professional qualifications to do so. Mass produced devices which need to be adapted to meet the specific requirements of the dental/medical practitioner or any other professional user are not considered to be custom made devices.

It has been generally accepted by the UK that dental appliances will be defined as custom made devices and the requirements of the Medical Devices Directive will therefore apply to those who wish to manufacture these products.

Particular requirements
The particular requirements of the Medical Devices Directive which relate to custom made devices are:
- conformity assessment procedures (these are referred to in the next section)
- registration of person responsible for placing devices on the market
- the manufacturer or his authorised representative must register with the Competent Authority of the Member State; the UK Competent Authority will issue a uniform registration application form
- dental appliances are not required to carry a CE marking (whereas new dental equipment, instruments and materials must carry the CE marking).

Before a device is placed on the market there is a statement drawn up by the manufacturer, or his authorised representative established in the Community, that the device conforms to the requirements of the Directive.

Further reading
EC Medical Devices Directive No 10 Guidance notes for manufacturers of dental appliances.

Devices for special purposes
In order to meet the conformity assessment procedures the manufacturer of a dental appliance must draw up a statement for each appliance which contains the following information:
- data allowing identification of the appliance in question
- a statement that the appliance is intended for exclusive use by a particular patient together with the name of the patient
- the name of the dental practitioner (or other authorised person) who made that prescription and, where applicable, the name of the clinic concerned
- the particular features of the appliance as specified in the relevant prescription
- a statement that the appliance in question conforms to the essential requirements and, where applicable, indicating which essential requirements have not been fully met together with the grounds.

The Directive requires that the manufacturer must identify the link between the patient, the dentist and the dental laboratory and that this must be defined for every appliance together with the particular features of the design as defined by the dentist.

Keeping records
In order to maintain adequate records of the construction of custom made devices it is considered that the manufacturer would need to fulfil the following requirements:
- a documented review of the dentists' requirements to ensure that adequate information has been supplied by the dentist and to demonstrate an understanding of the manufacturing requirements for the design ie choice of materials, processing parameters
- manufacturing under controlled conditions eg following defined/documented processes and have some method of demonstrating they are being followed (eg records); using suitably qualified personnel; where appropriate undertaking calibration and maintenance of equipment; considerations of cleanliness and infection control; defined handling activities and packaging
- a documented review of the final product against the dentist's initial requirements before it is returned to the dentist and fitted in the patient's mouth.

It is apparent that the records of the construction of all dental appliances will need to be maintained for a period of at least five years and it is also clear that the regulatory body has the right to audit the control system as and when it deems it necessary.

Labelling dental appliances
The minimum requirements regarding the labelling supplied to the patient with the dental appliance should include:
- the name or trade name and address of the manufacturer
- the details strictly necessary for the user to identify the device and the contents of
- the packaging
- the words 'custom made'
- any special storage and/or handling conditions
- any warnings and/or precautions to take.

Appendix 8: cross-infection control

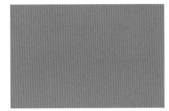

Media publicity has increased public awareness of the need for adequate and obvious cross infection control within the dental surgery. Patients now expect and demand high standards even though their knowledge of aspects of cross infection control is limited. General dental practitioners working within the regulations of the National Health Service must satisfy the Family Health Services Authority that they have adequate facilities in their practices for the sterilisation of dental equipment and that they comply with the current guidelines for cross infection control.

The British Dental Association published clear guidelines in 1986, which were subsequently updated in 1991 and 1996, on the necessity for strict cross-infection control for the safety of patients and operators. An additional document 'BDA Occasional Paper No 2 Infection Control in Dentistry: A practitioners guide', is also available as a practical restatement of advice given.

Dental personnel in general comply with guidelines for cross infection control to protect their patients, colleagues and themselves. They are aware that cross infection control is the responsibility of the entire dental team and that its overall effectiveness can be limited by the standard of any of its members as well as the degree to which they co-operate.

The possibility of the spread of infection or disease through the dental laboratory has been reported in the dental literature and methods of infection control have been evaluated. Transmissible microrganisms may enter the dental laboratory on dental appliances and impressions. A number of micro-organisms can be potentially involved in cross infection, for example; bacteria; (eg staphylococci, pseudomonas, mycobacteria), yeasts; (eg candida spp), viruses; (eg herpes, human immunodeficiency virus and hepatitis).

The extent to which commercial dental laboratories are used varies between dental practices however it is noteworthy that for the year 1996/1997 the Dental Practice Board for England and Wales recorded 24.7 million items of service claims and of these, approximately 4.5 million claims involved work by dental technicians.

It is a matter of concern that anyone today may call him or herself a dental technician, irrespective of whether, or to what extent, he or she has received officially recognised education and training. There is also no requirement for dental laboratories to be run by someone holding a recognised qualification in dental technology.

Dental technicians may be at a very slight increased risk of Hepatitis B and other infections from laboratory work received which has been in contact with patients' blood and saliva, although the latter vary in their degree of hazard of infection. Although the risk may be small, every effort should be taken, both within and beyond the dental surgery, to minimise it by following the current BDA guidelines.

The attitudes to cross infection control of 800 commercial dental laboratories registered with the Dental Laboratories Association has been surveyed and reported in the dental literature. Despite the topicality of the subject material and the need for careful cross infection control within and beyond the dental surgery, the response rate was low (22%). This survey, carried out in 1994, showed that cross infection control varied significantly between dental laboratories and similarly between dental practices, with many dental practitioners apparently sending undisinfected work to the dental laboratory:

- only 49% of respondents had a cross infection control policy
- of those with no policy, 64% claimed to intend to implement one in the future
- 30% of laboratories received known undisinfected work from the dental surgery
- the most popular chemicals used for disinfection are household bleach, chlorhexidine and glutaraldehyde
- of those items disinfected on arrival at the laboratory, those most frequently disinfected were dental impressions (77%) and dentures (51%). (The World Health Organisation, Center for Disease Control, (United States Department of Health Education and Welfare) and the British Dental Association in association with the Dental Laboratories Association and the British and American Dental Associations, have produced guidelines on the disinfection of dental appliances and dental impressions. It is suggested that appliances should be disinfected by immersion in a solution of sodium hypochlorite at a concentration of 10,000 parts per million available chlorine or a solution of proprietary disinfectant, suitable for dental use, at least as

complete dentures

effective as disinfection as 1:5/1:10 dilution of 5% to 10% solution of sodium hypochlorite. Problems disinfecting dental impressions have largely been overcome by the introduction of the use of chemicals such as sodium peroxymonosulphate, which is active against a broad spectrum of microrganisms but which at the same time does not have deleterious effects on the dimensional stability of widely used impression materials, such as alginate, polyethers and silicones)

- 44% of the respondents generally (90% or more of the time) wore gloves when handling dental work received and opened in the dental laboratory. (Gloves, if intact, form an effective barrier and their routine use is essential for good cross infection control)
- 74% wore protective eye spectacles when trimming or polishing dentures
- 61% used no disinfectant in the pumice and did not disinfect the polishing instruments, for example wheels and mops (micro-organisms have been cultured from pumice slurries)
- 46% had a policy for the immunisation of staff against Hepatitis B. (Hepatitis B is of concern to dental personnel and can be prevented by vaccination. Vaccination against Hepatitis B does not however protect against Hepatitis C, F or G. It is also recommended that dental staff are vaccinated against Diptheria, Hepatitis B, Pertussis, Poliomyelitis, Rubella, Tetanus and Tubercolosis).

Policies for infection control within hospital dental laboratories are in existence supporting recommendations made by the British Dental Association. A detailed policy for controlling cross infection within one dental laboratory was formally implemented in 1989. In brief, it consists of the disinfection of all technical work on the clinic prior to sending it to the dental laboratory. All work is rinsed with cold water to remove visible blood and saliva contamination and it is then spray disinfected with 0.5% chlorhexidine gluconate and put in a sealed clear polyethylene bag for dispatch to the dental laboratory. A piece of wet cotton wool is added to those impressions which need to be kept moist. They are not wrapped in a damp gauze as this makes subsequent inspection by the technician difficult and may conceal blood contamination.

All work is opened in a receiving area in the laboratory by technicians wearing gloves and the polythene bags are disposed of in a controlled way (fig. A8.1). The dental impressions are then spray disinfected using a Nebucid 840 disinfectant spray dispenser (Gimad, Italy, distributed by Unident SA, Geneva, Switzerland) using a broad spectrum anti-bacterial, anti-viral and anti-fungal agent, Dermacol (Unident SA, Geneva, Switzerland). Work leaving the laboratory is similarly washed and decontaminated and heat sealed in polythene bags.

Technicians wear protective spectacles when carrying out trimming and polishing procedures and the wheels and mops used are disinfected with 0.5% chlorhexidine gluconate daily. All dental technicians are given the opportunity to be immunised against Hepatitis B and a register maintained. Although effort was required initially to adapt to the changes in procedures in the dental laboratory, the technicians are satisfied with the extra protection they have been afforded.

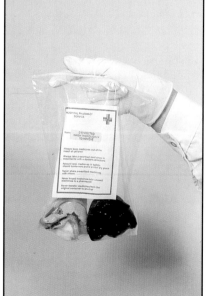

Fig. A8.1 Impressions have been dispatched to the dental laboratory in a sealed clearly labelled transparent plastic bag in order that they can be easily inspected by the technician. If there is evidence of contamination with blood the bag is returned unopened to the clinic for decontamination and disinfection.

Index

Abrasive cleaning pastes 17-18, 19
Acidic denture cleansers 18, 19
Acrylic dentures hygiene 19
Acrylic resin teeth 15
Aesthetic aspects 13-16
Alcohol abuse 22
Alkaline hypochlorites 18, 19
Allergy 7
Anterior teeth
    colour 14
    position 14-15
    selection 13-14
    size 14
Antidepressant drugs 23
Anxiety 21
    orofacial symptoms 23
Artificial teeth
    material 15
    personalisation of appearance 15-16
    selection 13
Atrophic mandibular ridge 4, 6-7, 41
Autopolymerising resin repairs 10-11

Behaviour therapy 23
Body dysmorphic disorder 23
Border moulding 2
Border seal 2
Brushes 17, 18
Bruxism 23

Chemical denture cleansers 18
Cleft palate 40, 41
Congenital oral maxillofacial defects 41
Copy dentures 4, 33
    technique 33-5
Cross-infection control 45-6
Cuspless teeth 3, 4, 31
    advantages/disadvantages 32
    setting up technique 31

Deformation 9-10
Dementias 22
Denture cleansers 17-18
    chemical 18
    ideal properties 17
    mechanical 17-18
Denture fixatives 4, 37-8
    constituents 37
    denture hygiene 19, 38
Denture hygiene 19-20
    cleaning advice 17
    soft lining materials 42
Denture phobias 23

Denture-induced stomatitis 18, 40
Depression
    neurotic 21
    orofacial symptoms 23
    psychotic 22
Disinfectant denture cleansers 18
Drug abuse 22

Effervescent peroxides 18
Ekbom's syndrome 23
Enzyme denture cleansers 18

Fibre-reinforced PMMA 11
Fit surface 5
Flabby maxillary ridge 4
Flangeless dentures 10
Fractures 9-12
    flexural fatigue 9
    impact 9
    predisposing factors 9-11
    permanent soft linings 41-2
    preventive reinforcement 11
    repair 11-12
    repeat 10-11
Fraenal attachments 4
Fraenal notch 10
Functional impressions 40

Gag reflex, heightened (retching) 2, 23, 25-6
    causes 25
    management 25-6

Hepatitis B 45, 46
High impact acrylic resin 10, 11, 42
Horseshoe design 10, 26
Hypnosis 25
Hypochondriacal neurosis 21-2
Hysterical neurosis 21

Immediate denture relines 40
Impression surfaces 1-2, 5

Labelling dental appliances 44
Lip biting 5
Locked occlusion 3, 6, 10
Loose dentures 1-4, 10
    impression surfaces 1-2
    neuromuscular disorders 4
    occlusal surfaces 3
    polished surfaces 2-3

Mandibular torus 7, 41
Mania 22

Manic depressive psychosis 22
Maxillary torus 7, 10, 41
Mechanical denture cleansers 17-18
Medical Devices Directive 43-4
Mental nerve trapping 7, 41
Metal reinforcements 11, 42
Methyl methacrylate monomer allergy 7
Midline diastema 10
Monosymptomatic hypochondriacal psychosis 22-3
Mylohyoid ridges 4, 7, 41

Neuromuscular disorders 4
Neuroses 21-2
    orofacial symptoms 23
Neutral zone 2-3, 4, 5, 27
    clinical technique 27-8, 29
Non-anatomic teeth see Cuspless teeth

Obsessional anxiety neurosis 21
Occlusal loading 10
Occlusal surfaces 3, 6
Occlusion adjustment 3, 6
Oral maxillofacial defects 41
Overextended borders 2, 5

Painful dentures 5-7, 23, 41, 42
    allergy 7
    impression surfaces 5
    occlusal surfaces 6
    polished surfaces 5
    psychological problems 7
    systemic disease 7
    unfavourable anatomy 6-7
    xerostomia 7
Palatal relief chamber 1, 2
Personalisation of appearance 15-16
Personality disorders 22
    orofacial symptoms 23
Phantom bite syndrome 23
Phobias 22
    denture 23
Polished surfaces 2-3, 5
Poly(methyl methacrylate) 9
    allergy 7
    reinforcement 11
Porcelain teeth 15
Post-dam 1-2, 5, 26
Premature occlusal contact 3, 6
Psychiatric disorders 7, 21-3
    classification 21-2
    clinical management 23
    orofacial manifestations 22-3

complete dentures

Psychoses 22
    orofacial symptoms 22-3

Records 44
Relief provision 5, 10, 41
Repairs 11-12
    autopolymerising resin 10-11
Rubber-reinforced PMMA 11

Saliva 4
Schizophrenia 22
    orofacial symptoms 23
Sedation 25
Soft lining materials 5, 7, 39-42
    cleansing 19, 40, 42
    functional impressions 40
    permanent 40-2
    adhesion 42
    adjustment 42
    temporary 39-40
    tissue conditioners 39-40
Stress concentration 10
Systemic disease 7

Temporary soft materials 39-40
    cleansing 19, 40
Temporomandibular joint dysfunction 23
Tissue conditioners 39-40
Tongue biting 5
Training plate 26

Ultrasonic cleaners 17, 18
Underextended flanges/borders 2, 10

Warped denture 1
Working models 1
Worn occlusion 10

Xerostomia 4, 7, 23, 41